Monographs in Electrical and Electronic Engineering

Series editors: P. Hammond and T. J. E. Miller

Monographs in Electrical and Electronic Engineering

Semiconductor devices, circuits, and systems

Albrecht Möschwitzer

Professor, Technical University Dresden

CLARENDON PRESS · OXFORD

1991

Oxford University Press, Walton Street, Oxford OX2 6DP

Oxford New York Toronto
Delhi Bombay Calcutta Madras Karachi
Petaling Jaya Singapore Hong Kong Tokyo
Nairobi Dar es Salaam Cape Town
Melbourne Auckland
and associated companies in
Berlin Ibadan

Oxford is a trade mark of Oxford University Press

Published in the United States
by Oxford University Press, New York

British Library Cataloguing in Publication Data
Möschwitzer, Albrecht
Semiconductor devices, circuits, and systems.
1. Semiconductor circuits
I. Title   II. Series
621.3815
ISBN 0-19-859374-0

Library of Congress Cataloging in Publication Data
Möschwitzer, Albrecht.
Semiconductor devices, circuits, and systems / Albrecht Möschwitzer
(Monographs in electrical and electronic engineering; 24)
1. Semiconductors.   2. Solid state electronics.   I. Title.   II. Series.
TK7871.85.M5754   1991   621.381'52 — dc20   90-25666
ISBN 0 19 859374 0

Typeset by Colset Pte Ltd, Singapore
Printed in Great Britain by
Bookcraft (Bath) Ltd
Midsomer Norton, Avon

To my father and mother,
and to my family

Preface

Modern electronics is about implementing hardware functions in semiconductor chips and about the software that runs these semiconductor circuits. Very large scale integration (VLSI) of electronic circuits and systems needs interdisciplinary work by device physicists, process developers, circuit designers, design automation specialists, and computer architects. Today's electronic engineers must have a background knowledge ranging from the device level through circuit level to the systems level. This is a challenge to modern curricula in electronic engineering and is a result of the rapid development of solid state (semiconductor) electronics. In 1950 a semiconductor component was just a single transistor, in 1960 a simple circuit (logic gate or flip-flop), and today it is a complex system (e.g. a CPU of a 32-bit mainframe computer or signal processor).

This book covers all these topics from semiconductor devices to systems in a compact manner. The text outlines the latest advances in semiconductor devices for VLSI circuits but also includes simple and easy to use analytical models as well as results of device simulation. The section on circuits gives an overview of basic bipolar and field-effect transistor gates and is mainly devoted to CMOS standard cells and functional blocks (macrocells) which are successfully applied in numerous large-scale integrated CMOS chips. In the future CMOS will be the most important circuit type. The section on systems outlines the top-down design style of digital systems (mainly processors and memories) using functional blocks described in the previous circuit section. Finally some problems of testing and details of physical layout of chips are considered. The importance of digital signal processing continues to grow with respect to analogue signal processing. Therefore we consider mainly digital circuits.

Exercises are included at the end of each chapter. As background to this text, introductory courses such as 'Electron Physics', 'Electronic Devices and Circuits', or 'Computer Engineering' would be helpful.

Special thanks are due to Professor Percy Hammond of the University of Southampton who encouraged me to write this book. I am also grateful to the staff of OUP for their patience and co-operation.

Dresden   A.M.
August 1990

Contents

List of symbols

A	area
A_N	common-base d.c. current gain
B_N	common-emitter d.c. current gain
c	light velocity
C	capacitance
C''	capacitance per unit area
E	electric field
e	$= 1.6 \times 10^{-19}$ As, electron charge
f	frequency
h	$= 6.625 \times 10^{-34}$ Js, Planck's constant
I	d.c. current
i	a.c. current
$\mathbf{I}$	complex phasor of current
k	$= 1.38 \times 10^{-23}$ J K^{-1}, Boltzmann constant
m	mass
n	electron density
n_i	intrinsic density
N_A	acceptor density
N_D	donor density
p	hole density
Q	charge
Q''	charge per unit area
R	resistance
r	a.c. resistance
S	current density
t	time
T	absolute temperature
V	voltage, potential (d.c.); v (a.c.); $\mathbf{V}$ (complex phasor)
V_{bi}	built-in potential
V_p	pinch-off voltage
V_t	threshold voltage
V_{AB}	voltage drop between A and B
V_{FB}	flat-band voltage
V_{FO}	floating potential
$V_T = \dfrac{kT}{e}$	thermal voltage
W	energy
W_{EA}	electron affinity

W_g	energy gap
φ	potential
ϵ	permittivity
κ	conductivity
μ	mobility
ρ	resistivity

1 Basic semiconductor structures

All semiconductor devices and integrated circuits are made with basic structures. These are doped semiconductor layers, p-n junctions, MOS sandwich structures, contacts, and interconnection lines. The operation of these structures is based on the generation, distribution, and flow of charge carriers in semiconductors, insulators, and metals. Therefore, in this chapter, we consider the fundamentals, construction, electronic behaviour, and basic application of these elementary structures.

1.1 CHARGE CARRIERS IN SEMICONDUCTORS

1.1.1 Electrons and holes in intrinsic and doped semiconductors

In semiconductors we have two types of charge carriers: electrons (mass $m_n \approx 10^{-30}$ kg, charge $e = -1.6 \times 10^{-19}$ As) and holes (mass $m_p \approx 10^{-30}$ kg, charge $e = +1.6 \times 10^{-19}$ As) (Shockley 1950; Yang 1978). These are generated in pairs by thermal excitation (Fig. 1.1(a)) of valence band electrons or by impurity ionization (Fig. 1.1(b), (c)), where an energy $W > W_C - W_D$ resp, $W > W_A - W_V$ is needed (for energy-band diagrams see 1.2.).

In a semiconductor that is doped with donors such as phosphorus, we get excess electrons in the conduction band (energy $W > W_C$; see Fig. 1.1(b)) if the donors are ionized. If the semiconductor is doped with acceptors such as boron we get defect electrons (holes) in the valence band (energy $W < W_V$; see Fig. 1.1(c)) if the acceptors are ionized. Normally at room temperature all donors (concentration N_D) and acceptors (concentration N_A) are ionized, $N_D^+ = N_D$, $N_A^- = N_A$.

In a semiconductor having only donors the electron concentration is therefore approximately $n \approx N_D$, where N_D is the donor concentration and in a semiconductor having only acceptors the hole concentration is approximately $p \approx N_A$, where N_A in the acceptor concentration. (For more details see Section 1.1.3..)

1.1.2 Energy-band diagrams

To explain the operation of semiconductor devices, energy-band diagrams are very helpful. These are plots of energy and potential levels against the location as shown in Fig. 1.2. First we see the energy levels of the conduction band edge, W_C, and valence band edge, W_V. The energy gap between W_V

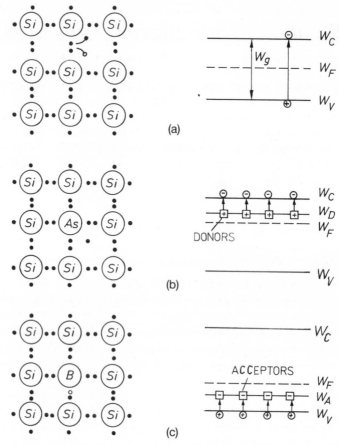

Fig. 1.1. Two-dimensional schematic representation of the crystal structure and energy-band diagram in (a) an intrinsic semiconductor, (b) a semiconductor doped with donors, and (c) a semiconductor doped with acceptors.

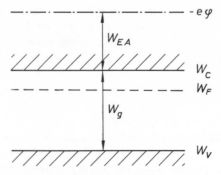

Fig. 1.2. Energy-band diagram.

and W_C is the forbidden gap, W_g. This means that electrons in a solid can have only restricted values of energy. The energy required to free (emit) an electron from the conduction band edge W_C is the electron affinity, W_{EA}. For silicon we have $W_{EA} = 4.1\,\text{eV}$ for emission into vacuum and $W_{EA} = 3.25\,\text{eV}$ for emission into SiO_2. The forbidden gap at room temperature is $W_g = 1.1\,\text{eV}$. This energy-band diagram has already been

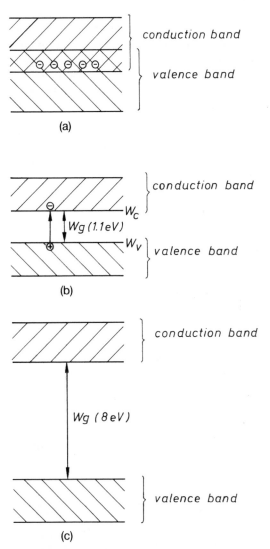

Fig. 1.3. Energy-band diagram of (a) a metal, (b) a semiconductor, and (c) an insulator.

used in Fig. 1.1 to explain the generation of electrons and holes in intrinsic (Fig. 1.1(a)) and doped semiconductors (Fig. 1.1(b), (c)).

In Fig. 1.3 we compare energy-band diagrams for a metal, a semiconductor, and an insulator. In a metal we have no forbidden gap and the valence band overlaps the conduction band. This means all electrons contribute to electrical conduction. An insulator has a large forbidden gap. In this case a high energy is necessary to generate electrons and holes. At room temperature this is very unlikely. So we have almost no mobile charge carriers and no electrical conductivity in the insulator.

The forbidden gap, W_g, can be altered in compound semiconductors as shown in Fig. 1.4. In this figure W_g is plotted as a function of the mixture ratio x. These compound semiconductor materials are used for optoelectronic and high-speed devices (Shur 1987). Heterostructures which have very interesting electronic behaviour can also be made with these compounds (Milnes 1986).

1.1.3 Temperature dependence of charge carrier densities

The charge carrier densities are temperature dependent. For an n-type semiconductor which is doped with N_D donors we have a typical

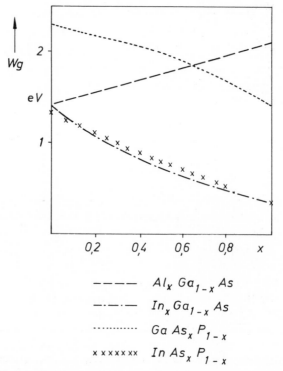

Fig. 1.4. Band-gap width for compound semiconductors.

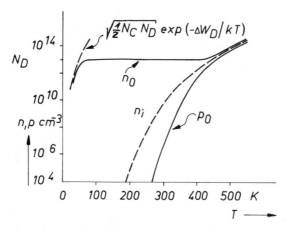

Fig. 1.5. Temperature dependence of electron and hole concentrations.

temperature dependence of the electron density, n, as shown in Fig. 1.5. At low temperatures only a few donors are ionized. The electron density n_o is small and increases exponentially with temperature according to (Möschwitzer and Lunze 1990)

$$n_o(T) = \sqrt{\tfrac{1}{2}N_C N_D}\, \exp\left(-\frac{\Delta W_D}{2\,kT}\right).$$ (1.1)

$\Delta W_D = W_C - W_D$ is the activation energy of the donors (see Fig. 1.1(b)) and has a typical value of $\Delta W_D = 0.05$ eV, N_C is the effective density of states in the conduction band (Shockley 1950). Typically $N_C \approx 10^{19}\,\mathrm{cm}^{-3}$; see Equation (1.6a), k is Boltzmann's constant and T is the absolute temperature.

At medium (room) temperatures all donors are ionized ($N_D = N_D^+$) and we have

$$n_o \approx N_D.$$ (1.2)

At high temperatures electrons that are directly generated from the valence band (see Fig. 1.1(a)) give a further increase in the total electron density

$$n_o = \sqrt{N_C N_V}\, \exp\left(-\frac{W_g}{2\,kT}\right) = p_o$$ (1.3)

where electron (n_o) and hole densities (p_o) are equal. N_V is the effective density of states in the valence band (see Equation (1.6b)).

The ionization ratio N_D^+/N_D of the donors depends not only on the temperature but also on the donor concentration. This is shown in Fig. 1.6 (Overstraeten *et al.* 1987). From this figure we can see that the ionization ratio N_D^+/N_D is decreased at high donor concentration N_D. This effect is called impurity freeze-out.

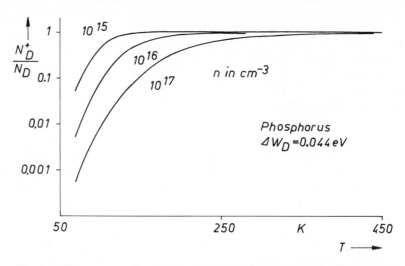

Fig. 1.6. Temperature dependence of ionized donors (impurity freeze-out).

If we dope an n-type semiconductor additionally with acceptors of a concentration $N_A < N_D$ then, in the n-type semiconductor with fully ionized donors and acceptors, we obtain a total electron density of

$$n_o = N_D - N_A. \tag{1.4}$$

1.1.4 Energy distribution of charge carriers

The energy distribution of charge carrier densities is determined by the Fermi–Dirac statistics. For non-degenerate semiconductors ($n_o < N_C$, $p_o < N_V$) we derive, for the electron density in thermal equilibrium (Shockley 1950),

$$n_o = N_C \exp\left(-\frac{W_C - W_F}{kT}\right), \tag{1.5a}$$

and for the hole density

$$p_o = N_V \exp\left(-\frac{W_F - W_V}{kT}\right), \tag{1.5b}$$

where W_F is the Fermi level. This is the energy level at which the occupation probability of an energy state with electrons is $\frac{1}{2}$. N_C and N_V are the effective density of states in the conduction band and valence band, respectively (Shockley, 1950):

$$N_C = 2\left(\frac{2\pi m_n kT}{h^2}\right)^{3/2} \approx 10^{19}\,\mathrm{cm}^{-3} \tag{1.6a}$$

$$N_V = 2 \left(\frac{2\pi m_p kT}{h^2} \right)^{3/2} \approx 10^{19} \, \text{cm}^{-3} \qquad (1.6b)$$

where h is Planck's constant.

A semiconductor is called non-degenerate if the electron and hole concentrations are smaller than the effective density of states ($n_o < N_C$, $p_o < N_V$). The product of Equations (1.5a) and (1.5b) yields

$$n_i^2 = n_o p_o = N_C N_V \exp\left(-\frac{W_g}{kT}\right). \qquad (1.7)$$

n_i is the intrinsic density and has a strong temperature dependence. For example, in silicon at $T = 300\,\text{K}$ it is $n_i = 1.5 \times 10^{10} \, \text{cm}^{-3}$. $n_i(T)$ is plotted in Fig. 1.5.

In an n-type semiconductor we have $p_{on} = n_{on}/n_i^2 \leqslant n_{on}$. Therefore in an n-type semiconductor electrons are majority carriers and holes are minority carriers. In p-type semiconductors holes are the majority carriers and electrons are the minority carriers.

1.1.5 Impact ionization (avalanche effect)

So far we have considered only thermal generation of charge carriers, but there are other mechanisms for charge carrier generation in semiconductors. First let us consider carrier generation by impact ionization. In high electric fields, electrons and holes are accelerated to such high velocities that an impact with the atoms of the semiconductor grid (see Fig. 1.1) causes ionization and frees electron–hole pairs. The number of electron–hole pairs generated by impact ionization is proportional to the densities n, p and the velocities v_n, v_p of electrons and holes involved and the ionization coefficients, $\alpha_n(E)$, $\alpha_p(E)$. The generation rate (generated electron–hole pairs per unit time and volume) is therefore

$$G = n v_n \alpha_n(E) + p v_p \alpha_p(E). \qquad (1.8)$$

The ionization coefficients for electrons $\alpha_n(E)$ and holes $\alpha_p(E)$ are strongly (exponential) dependent on the electrical field as shown in Fig. 1.7 for silicon (Si), germanium (Ge) and galliumarsenide (GaAs).

1.1.6 Photogeneration

If a photon with an energy $W = hf$ is absorbed in a semiconductor, charge carriers can be generated. For the generation of electron–hole pairs it is necessary for the energy of the photons to be larger than the energy gap ($hf > W_g$). The generation rate of an electron–hole pair at a depth x measured from the surface is

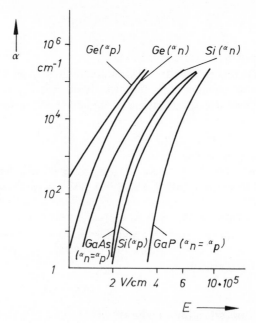

Fig. 1.7. Dependence of the impact ionization coefficient on the electric field.

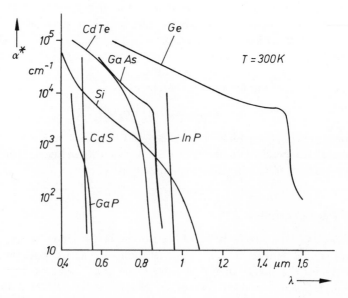

Fig. 1.8. Dependence of the absorption constant on the light wavelength.

$$G = \alpha^*(\lambda)\,\Phi_o\,\exp\{-\alpha^*(\lambda)x\}, \qquad (1.9)$$

where Φ_o is the photon flux density (cm^{-2}s^{-1}) at the surface. The absorption constant $\alpha^*(\lambda)$ is dependent on the wavelength λ of the incident light. For various semiconductors it is plotted against λ in Fig. 1.8. From Fig. 1.8 we see a rapid decrease of the absorption constant at high wavelength λ. It is zero at a wavelength $\lambda_g = hc/W_g$ which is high for semiconductors with small forbidden gap W_g.

1.1.7 Recombination

The opposite mechanism to carrier generation is recombination, as shown in Fig. 1.9. For direct recombination an electron moves from the conduction band directly to the valence band losing an amount of energy W_g. This energy can be conserved by emitting a photon (recombination radiation). The recombination rate for direct recombination is

$$R_D = \sigma_c\,v_{th}\,(pn - n_i^2), \qquad (1.10a)$$

where σ_c is the capture cross-section and v_{th} is the thermal velocity ($\approx 10^{+7}$ cm/s for silicon at room temperature; see Equation (1.113)).

Energy states in the forbidden gap can act as traps or recombination centres. Electrons and holes can be captured by these traps. Under steady state conditions, if the electrons and holes are trapped at the same rate by a trap, we have an indirect recombination via that trap (see Fig. 1.9). The recombination rate for indirect recombination is (Shockley and Read 1952)

$$R_1 = \frac{pn - n_i^2}{\tau_{po}(n + n_1) + \tau_{no}(p + p_1)} \qquad (1.10b)$$

where

$$n_1 = N_C \exp\left(-\frac{W_C - W_t}{kT}\right) \qquad (1.11a)$$

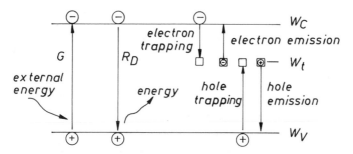

Fig. 1.9. Recombination–generation mechanisms in semiconductors.

$$p_1 = N_V \exp\left(-\frac{W_t - W_v}{kT}\right) \tag{1.11b}$$

and the life times

$$\tau_{po} = \frac{1}{N_t c_p} \tag{1.11c}$$

$$\tau_{no} = \frac{1}{N_t c_n} \tag{1.11d}$$

and capture coefficients

$$c_p = v_{th} \sigma_{cp} \tag{1.11e}$$

$$c_n = v_{th} \sigma_{cn} \tag{1.11f}$$

N_t is the total density of traps, σ_c is the capture cross-section and v_{th} is the thermal velocity (see Equation (1.113)).

If the deviations of the electron and hole densities, n and p, from their equilibrium densities, n_o and p_o, are very small we obtain, with Equation(1.10), the simplified formulae for low injection

$$R = \frac{n - n_o}{\tau_n} \tag{1.12a}$$

$$R = \frac{p - p_o}{\tau_p} \tag{1.12b}$$

where τ_n and τ_p are the electron and hole lifetimes, respectively.

1.1.8 Current flow

The current in semiconductors is due to electrons and holes moving with velocities v_n and v_p, respectively, and is given by

$$I = eA(nv_n + pv_p) \tag{1.13}$$

where A is the area.

At low electric fields the velocities are proportional to the field strength E:

$$v_n = \mu_n E \tag{1.14}$$

$$v_p = \mu_p E. \tag{1.15}$$

The mobilities μ_n and μ_p for electrons and holes are dependent on the impurity concentration N_B and on the temperature T as shown in Figs. 1.10 and 1.11. Because of the scattering mechanisms the mobilities decrease at high electric fields. Therefore the electron velocity in silicon saturates as

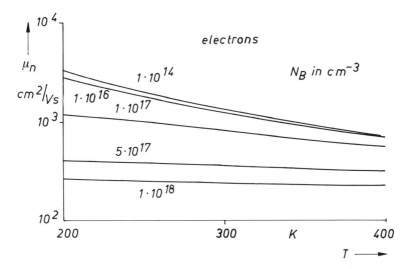

Fig. 1.10. Temperature dependence of the electron mobility in silicon.

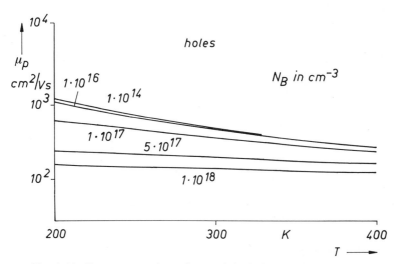

Fig. 1.11. Temperature dependence of the hole mobility in silicon.

shown in Fig. 1.12. For the some compound semiconductors, such as GaAs, the electron velocity can decrease with increasing electric field (see Fig. 1.13). This means a negative differential mobility ($\mathrm{d}v/\mathrm{d}E < 0$) and causes instabilities such as the Gunn effect (Shur 1987).

If we apply a step of high electric field the velocity is initially high. After a relaxation time the carrier scattering leads to a lower steady state velocity.

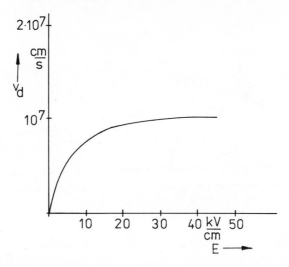

Fig. 1.12. Field dependence of electron drift velocity in silicon.

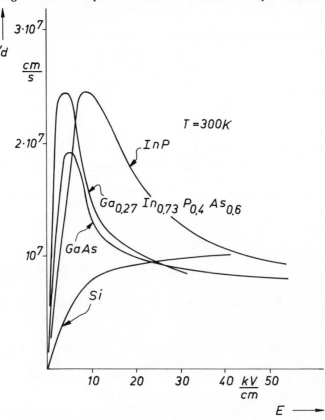

Fig. 1.13. Dependence of the electron drift velocity on the electric field in different semiconductors.

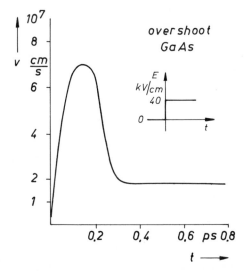

Fig. 1.14. Overshoot effect in GaAs.

This effect, which is illustrated in Fig. 1.14, is called overshoot and it can be used in very high-speed devices (Shur 1987).

1.1.9 The basic equations describing the macroscopic behaviour of semiconductor devices

Most of the analytical approximations and numerical simulations of semiconductor devices today are based on the following five equations for holes and electrons in semiconductors (Cham 1986; Selberherr 1984; Snowden 1989). First, we have the transport equations for electron and hole current densities S_n and S_p.

$$S_n = en\mu_n \mathbf{E} + eD_n \operatorname{grad} n \qquad (1.16)$$

$$S_p = ep\mu_p \mathbf{E} - eD_p \operatorname{grad} p. \qquad (1.17)$$

The first terms in Equations (1.16) and (1.17) are drift currents which have already been described in Section 1.1.8. The second terms are diffusion currents. D_n and D_p are the diffusion coefficients for electrons and holes, respectively. For medium-doped silicon we have, for example, $D_n \approx 20\,\text{cm}^2/\text{s}$ and $D_p \approx 10\,\text{cm}^2/\text{s}$. $\mathbf{E}$ is the electric field.
    In general,

$$D_{n|p}/\mu_{n|p} = kT/e = V_T.$$

Next we have the continuity equations

$$\operatorname{div} S_n = e(R - G) + \frac{\partial n}{\partial t} \qquad (1.18)$$

$$- \operatorname{div} \mathbf{S}_{p} = e(R - G) + \frac{\partial p}{\partial t}. \tag{1.19}$$

The recombination rates R and generation rates G can be derived from the results of Sections 1.1.5–1.1.7.

Finally we have Poisson's equation

$$\operatorname{div} \mathbf{E} = \frac{e}{\epsilon_{H}} (p - n + N_{D}^{+} - N_{A}^{-}) \tag{1.20}$$

which describes the interdependence of electric field and space charge. ϵ_{H} is the permittivity of the semiconductor.

1.2   RESISTORS

There are several types of resistors used in integrated circuits. First we have resistors made with doped semiconductor layers (Fig. 1.15). These resistors can be designed so that the circuit functions properly, or they may appear as parasitic elements. In any case we must know about them and calculate their effects.

The resistance of a resistor sheet of width b, length L, and depth d is

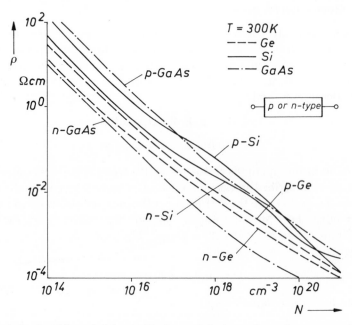

Fig. 1.15. Specific resistivity in Si, Ge, and GaAs vs. impurity concentration.

$$R = R_s \frac{L}{b} \qquad (1.21)$$

with the sheet resistance R_s is given by

$$R_s = \frac{1}{\kappa d}. \qquad (1.22)$$

The conductivity is

$$\kappa = e(\mu_n n + \mu_p p) \qquad (1.23)$$

and is dependent on the impurity concentration N in the resistor sheet. In Fig. 1.15 we show the resistivity $\rho = 1/\kappa$ for several n and p-type semiconductors (Sze and Irvin 1968). If we consider impurity freeze-out (see Section 1.1.3.) we get a temperature dependence of the conductivity at low temperatures as shown in Fig. 1.16.

In integrated circuits such resistor sheets are doped by diffusion or ion implantation and we obtain an inhomogeneous impurity profile as shown in Fig. 1.17. The impurity concentration, $N(x)$, and thus the conductivity, decreases with the distance, x, from the semiconductor surface. The mean conductivity, $\bar{\kappa}$, and the resistance, R, of an inhomogeneously doped sheet between depths x_A and x_B (see Fig. 1.17) are

$$\bar{\kappa} = \frac{1}{x_B - x_A} \int_{x_A}^{x_B} \kappa(x)\, dx \qquad (1.24)$$

and

$$R = \frac{1}{\bar{\kappa}(x_B - x_A)} \frac{L}{b} = R_s \frac{L}{b}. \qquad (1.25)$$

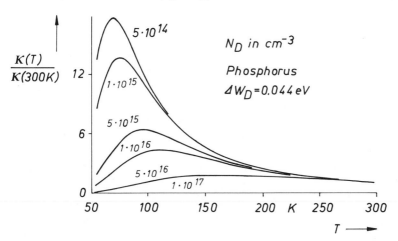

Fig. 1.16. Conductivity in an n-type semiconductor at low temperatures.

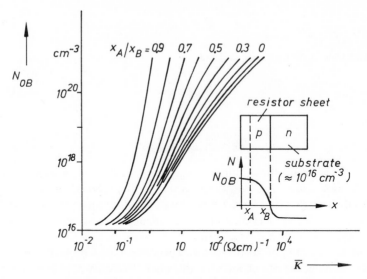

Fig. 1.17. Dependence of the mean conductivity in a doped p-type semiconductor sheet on the surface concentration of a Gaussian impurity distribution and the relative depth of the sheet ($N_B = 10^{16}\,\text{cm}^{-3}$).

Using an impurity profile (see Fig. 1.17) with a diffusion length L_D

$$N(x) = N_{OB} \exp\left(-x^2/L_D^2\right) - N_B \tag{1.26}$$

$\bar{\kappa}$ was calculated with $\kappa(x) = 1/\rho$ from Fig. 1.15 and the result is plotted in Fig. 1.17 for a substrate doping of $N_B = 10^{16}\,\text{cm}^{-3}$. The depth of the resistor sheet can be derived if we set $N(x_B) = 0$ in Equation (1.26):

$$x_B = L_D \sqrt{\ln \frac{N_{OB}}{N_B}} \; . \tag{1.27}$$

For unburied resistors $x_A = 0$.

The sheet resistance R_s (the resistance of a square resistor $b = L$ in Equation (1.21)) is an important design parameter. For p-type resistor sheets it is plotted as a function of the surface concentration of the impurities, N_{OB}, in Fig. 1.18. For this plot we used $x_A = 0$ and a substrate doping of $N_B = 10^{16}\,\text{cm}^{-3}$.

Fig. 1.19 shows a layout and the cross-section of an integrated resistor. From this drawing we see all the parasitic elements associated with the structure. Owing to the reverse biased p-n junction (see Section 1.4) we have a parasitic capacitance, C_s, to the substrate. It is modelled by a simple T-circuit as shown in Fig. 1.19. The contact region is modelled by a ladder R-network and we obtain, for the contact resistance,

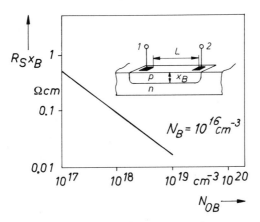

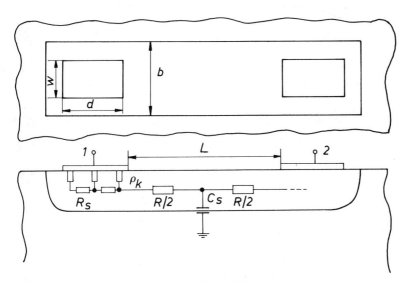

Fig. 1.18. Dependence of the sheet resistance of a doped p-type semiconductor sheet
on the surface concentration with Gaussian impurity distribution ($N_B = 10^{16}$ cm^{-3}

Fig. 1.19. Layout, cross-section, and model of a semiconductor resistor.

$$R_k = \sqrt{\rho_k R_s} \, \frac{1}{w} \coth \sqrt{\frac{R_s}{\rho_k}} \, d. \tag{1.28}$$

A typical value of the contact resistivity, ρ_k, of aluminium–p-type silicon is
$\rho_k = 1000 \ \Omega \, \mu m^2$. The total resistance, including the contact, is then

$$R = 2R_k + R_s \frac{L}{b}. \tag{1.29}$$

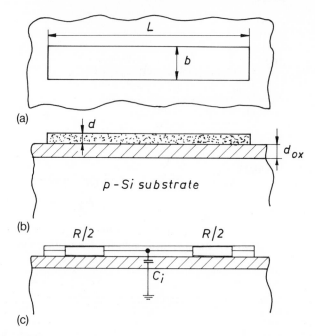

Fig. 1.20. Layout, cross-section, and model of a thin-film resistor.

Another resistor type is the thin-film resistor as depicted in Fig. 1.20. It occurs as a polysilicon, silicide, (e.g. WSi_2) and metal (e.g. Al) layer. In Table 1.1. some typical sheet resistances, R_s, are summarized. These elements are isolated from the substrate by thin dielectric layers, such as SiO_2. This gives a capacitive coupling to the substrate, as shown in Fig. 1.20(c), which can be modelled by a simple T-equivalent circuit.

1.3   MOS AND INSULATION-LAYER CAPACITORS

Two insulation-layer capacitors are sketched in Fig. 1.21. The structure in Fig. 1.21(a) represents a metal, polysilicon, or silicide layer running above

Table 1.1. Sheet resistances of various thin-film resistors.

material	$R_S (\Omega)$
Al	0.03
WSi_2	1–3
$MoSi_2$	1–3
Poly-Si (undoped)	$10^5 - 10^7$
n^+ Poly-Si	20–100

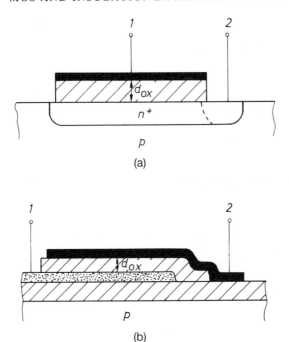

Fig. 1.21. Cross-section of (a) a MOS capacitor and (b) a thin-film capacitor.

a heavily-doped semiconductor substrate. The structure in Fig. 1.21(b) occurs in multilayer systems between polysilicon, silicide, and metal layers. The capacitance per unit area is

$$C''_{ox} = \epsilon_{ox}/d_{ox} \tag{1.30}$$

where ϵ_{ox} is the permittivity (e.g. for SiO_2 $\epsilon_{ox} = 3 \times 10^{-13}$ As/Vcm) and d_{ox} is the thickness of the dielectric layer.

MOS capacitors appear between a metal or polysilicon gate and a low-doped semiconductor substrate (Fig. 1.22). The capacitance of a MOS capacitor is voltage dependent. As shown in Fig. 1.22 the electric field penetrates into the low-doped semiconductor and creates a space-charge layer at the surface. The charge Q_R in the space-charge layer varies with the voltage drop across the semiconductor V_s. Owing to this voltage dependence we have a space-charge layer capacitance, per unit area, of

$$C''_R = \frac{dQ''_R}{dV_s}. \tag{1.31}$$

C_R is connected in series with the insulation-layer capacitor, C_i (Fig. 1.23(a)) and

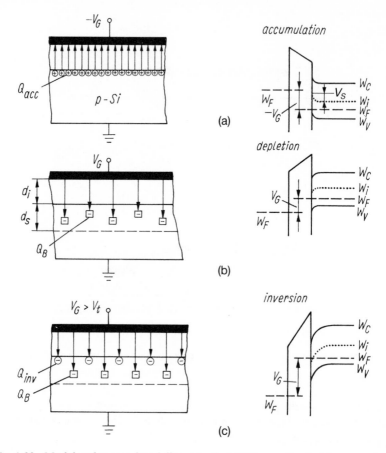

Fig. 1.22. Model and energy-band diagram of a MOS capacitor in (a) accumulation, (b) depletion, and (c) inversion.

$$C_i'' = \frac{\epsilon_i}{d_i} \tag{1.32}$$

where ϵ_i is the permittivity and d_i is the thickness of the insulation layer between the gate (metal or polysilicon) and the semiconductor. The MOS capacitance, per unit area, is therefore

$$C_M'' = \frac{C_i'' C_R''(V_s)}{C_i'' + C_R''(V_s)} \tag{1.33}$$

where C_R'' is voltage dependent. This will now be discussed in greater detail.

Let us consider a p-type semiconductor as an example. The space charge consists of three parts. The first part is the bulk charge of ionized acceptors (Möschwitzer and Lunze 1990)

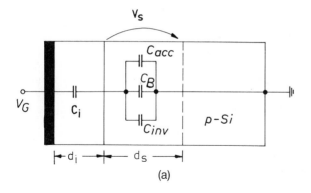

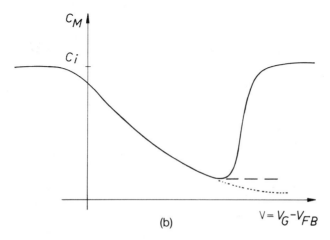

Fig. 1.23. MOS capacitor: (a) circuit model and (b) voltage dependence of the capacitance.

$$Q_B'' = eN_A d_s = \sqrt{2eN_A \epsilon_H V_s} \qquad (1.34)$$

where N_A is the acceptor concentration and d_s is the space-charge layer width, and the bulk-charge capacitance is (see Fig. 1.23)

$$C_B'' = \frac{dQ_B''}{dV_s} = \sqrt{\frac{e\epsilon_H N_A}{2 V_s}} \qquad (1.35)$$

where ϵ_H is the permittivity of the semiconductor (e.g. for silicon $\epsilon_H = 10^{-12}$ As/Vcm) and N_A is the acceptor concentration in the p-type semiconductor.

The second part is the hole accumulation charge, Q_{acc}'', appearing for negative gate voltages (or, more precisely, at negative surface potentials V_s as shown in Fig. 1.22(a)). In that case Q_R'' is strongly dependent on the voltage and therefore the capacitance is very high:

$$C''_{acc} = \frac{dQ''_{acc}}{dV_s} \gg C''_i. \tag{1.36}$$

The third part is the electron inversion charge, Q_{inv}, appearing at high positive gate biases ($V_G > V_t$, where V_t is the threshold voltage for strong inversion as shown in Fig. 1.22(c)). In this case Q_R is strongly dependent on the voltage and therefore the capacitance is again very high:

$$C''_{inv} = \frac{dQ''_{inv}}{dV_s} \gg C''_i. \tag{1.37}$$

We can now discuss the voltage dependence of the total MOS capacitance C_M using Equations (1.33)–(1.37) and Fig. 1.23. (Nicollian and Goetzberger 1967).

At negative gate biases (the accumulation case of Fig. 1.22(a)) we have $C''_R \approx C''_{acc} \gg C''_i$, by Equation (1.36), and therefore

$$C''_M \approx C''_i = \frac{\epsilon_i}{d_i}. \tag{1.38}$$

At low positive gate biases we have depletion (Fig. 1.22(b)), so $C''_R \approx C_B$ and we therefore have

$$C''_M = \frac{C''_i C''_B}{C''_i + C''_B} = \frac{\epsilon_i}{d_i} \left(1 + \frac{\epsilon_i}{d_i} \sqrt{\frac{2V_s}{e\epsilon_H N_A}}\right)^{-1}. \tag{1.39}$$

At high positive gate biases ($V_G > V_t$) we have strong inversion (Fig. 1.22(c)), where $C''_R \approx C''_{inv} \gg C''_i$ by Equation (1.37), and therefore we again have

$$C''_M \approx C''_i = \frac{\epsilon_i}{d_i}.$$

Finally we describe the relation between the external gate voltage, V_G, and the internal surface potential, V_s (Nicollian and Goetzberger 1967). Numerical results are plotted in Fig. 1.24. The surface potential V_s is zero at a negative gate voltage $V_G = V_{FB} < 0$. V_{FB} is called the flat-band voltage. V_{FB} may be considered to be a constant depending on the physical parameters of the MOS structure (e.g. contact potential, interface charge, etc.). In Chapter 2 we will consider this flat-band voltage in detail. For silicon-gate technology a typical value is $V_{FB} = -1$ V (Yang 1978). For low positive gate voltages (depletion) we have $V_s \sim V_G - V_{FB}$ (see Fig. 1.24). At high positive gate biases (strong inversion) we have

$$V_s \approx 2V_F = 2\frac{kT}{e} \ln \frac{N_A}{n_i} \tag{1.40}$$

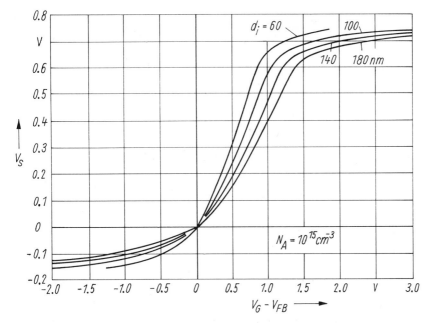

Fig. 1.24. Dependence of the surface potential on the gate voltage of a MOS capacitor.

(see also Fig. 1.24). Numerical results of C_M'' versus $V_G - V_{FB}$ are plotted in Fig. 1.25.

If we measure the MOS capacitance C_M at high frequencies we do not observe the increase of C_M in the inversion region. The measured capacitance at high frequencies is shown in Fig. 1.25 by the dashed lines. This effect can be explained as follows: rapid changes of the gate bias and surface potential do not yield rapid changes of the inversion charge, Q_{inv}. Therefore the capacitance due to the inversion charge C_{inv} is zero. The inversion charges Q_{inv} are minority carriers which cannot respond rapidly. It takes a long time (usually in the millisecond range) for a complete inversion layer to be built up at the surface. Minority carriers can move to the surface by a very small diffusion current or they can be generated at a small rate at the surface. Therefore the inversion layer charge can respond only very slowly to any change of the gate bias.

If we have no inversion layer, then even at high positive gate biases we obtain a deep depletion with a further increase of the space-charge layer width d_s and a further decrease of the MOS capacitance. In this case

$$C_M'' \approx C_B'' = \sqrt{\frac{e\epsilon_H N_A}{2V_s}} \qquad (1.41)$$

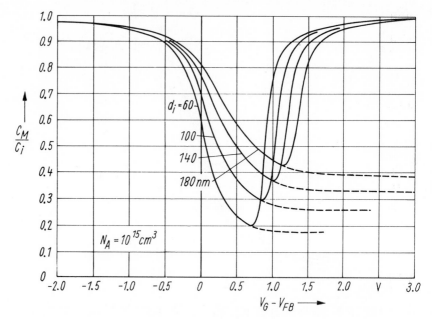

Fig. 1.25. Dependence of the M OS capacitance on the gate voltage.

(dotted lines in Fig. 1.23(b)), Deep depletion is applied in charge-coupled devices (CCD) (Sequin and Tompsett 1975).

The voltage dependence of the MOS capacitance can be used to build varactor diodes and for the investigation of the MOS structure itself (Möschwitzer and Lunze 1990; Muller and Kamins 1977; Nicollian and Goetzberger 1967; Yang 1978). MOS structures are basic elements from which field-effect transistors and integrated circuits are built. Therefore we will discuss these structures in greater detail in Section 2.2.

1.4   p-n JUNCTIONS AND p-n JUNCTION CAPACITORS

1.4.1   Introduction

The p-n junction is the most important elementary structure used in semiconductor devices and microelectronics (Shockley 1950). A p-n junction is made from p- and n-type semiconductor layers in the same crystal (Fig. 1.26).

Diffusion and ion implantation are the most important processes used in the making of junctions (Muller and Kamins 1977; Yang 1978). If we diffuse acceptor atoms in an n-type semiconductor doped with $N_D = N_B$ donors per unit volume we obtain an impurity distribution according to (see Fig. 1.26(b))

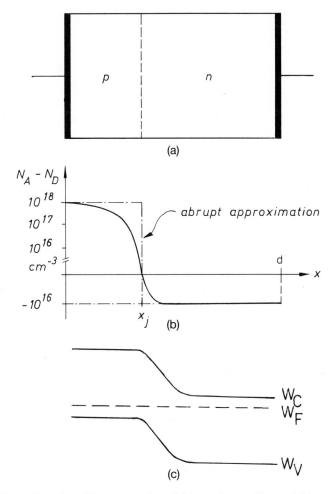

Fig. 1.26. p-n junction: (a) cross-section, (b) impurity profile, and (c) energy-band diagram.

$$N(x) = N_A - N_D = N_{OB} \exp\left(-\frac{x^2}{L_D^2}\right) - N_B \qquad (1.42)$$

where $L_D = \sqrt{4Dt}$ is a diffusion length which depends on the diffusion time t and the diffusion coefficient $D(T)$. $N_{OB} = Q_A / \sqrt{\pi Dt}$ is the surface concentration of the acceptors which depends on the number of acceptor atoms, Q_A, deposited on the surface per unit area. For ion implantation we also have a Gaussian distribution of the impurities

$$N(x) = N_A - N_D = N_m \exp\left(-\frac{(x - x_m)^2}{2\Delta x_m^2}\right) - N_B \qquad (1.43)$$

where x_m is the mean range and Δx_m is the standard deviation (Yang 1978).

We obtain the location of the p-n junction, x_j, from Equations (1.42) or (1.43) with $N(x_j) = 0$:

$$x_j = L_D \sqrt{\ln \frac{N_{OB}}{N_B}} . \tag{1.44}$$

As usual, throughout this book we use an abrupt approximation to the impurity profile (see Fig. 1.26(b)):

$$N = N_{OB} = N_A, \qquad 0 \le x < x_j \tag{1.45a}$$

$$N = -N_B = -N_D, \qquad x_j < x \le d. \tag{1.45b}$$

This gives good qualitative results and considerably simplifies all the following analytical estimations. The energy-band diagram of a p-n junction is sketched in Fig. 1.26(c).

1.4.2 Space-charge layer

The electron and hole distribution at the p-n junction is shown in Fig. 1.27(b). In the neutral p- and n-regions the majority carriers are equal to the impurity concentrations: $p_{op} = N_A$ and $n_{on} = N_D$; and the minority carrier densities are much smaller: $p_{on} = n_i^2/N_D$ and $n_{op} = n_i^2/N_A$. But near the junction ($x = 0$) electrons diffuse into the p-region and holes diffuse into the n-region. Owing to this diffusion of electrons and holes the ionized impurities near the junction are no longer neutralized and we have a space-charge layer in $-x_p \le x \le x_n$ of width $d_s = x_p + x_n$. The space charge in this layer is determined mainly by the ionized acceptors ($-eN_A$) and the ionized donors ($+eN_D$), as shown in Fig. 1.27(c). The space charge is

$$\rho = -eN_A \qquad -x_p \le x < 0, \tag{1.46a}$$

$$\rho = +eN_D \qquad 0 < x \le x_n. \tag{1.46b}$$

Because of this space charge we have an electric field $E(x)$ (Fig. 1.27(d)) which can be derived by solving Poisson's equation (1.20)

$$\frac{dE}{dx} = \frac{e}{\epsilon_H}(N_D^+ - N_A^-). \tag{1.47}$$

The maximum field in the centre of the p-n junction is given by

$$E_{max} = \frac{eN_A}{\epsilon_H}x_p = \frac{eN_D}{\epsilon_H}x_n \tag{1.48}$$

and the neutrality condition yields

$$x_p N_A = x_n N_D. \tag{1.49}$$

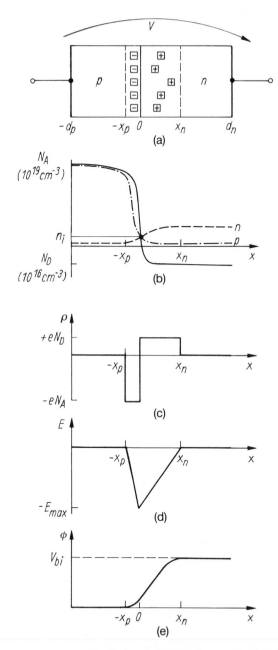

Fig. 1.27. p-n junction: (a) cross-section showing impurity space charge, (b) distribution of electron and hole concentration, (c) space-charge distribution, (d) electric field, and (e) potential.

The electric field creates a drift current which is equal to the diffusion current in thermal equilibrium so there is no resultant current flow in thermal equilibrium.

A further integration of Poisson's equation.

$$-\frac{d^2\phi}{dx^2} = \frac{dE}{dx}$$ (1.50)

gives the potential distribution $\phi(x)$ (Fig. 1.27(e)). Using the boundary conditions $E(-x_p) = E(x_n) = 0$; $E(-0) = E(+0)$; $\phi(-0) = \phi(+0)$, $\phi(-x_p) = 0$; $\phi(x_n) = V_{bi}$ we obtain, for thermal equilibrium (no external bias) (Möschwitzer and Lunze 1990)

$$V_{bi} = \frac{kT}{e} \ln \frac{N_A N_D}{n_i^2} = \frac{kT}{e} \ln \frac{n_{on} p_{op}}{n_i^2}$$ (1.51)

$$x_n = \sqrt{\frac{2\epsilon_H N_A V_{bi}}{eN_D(N_A + N_D)}}$$ (1.52)

$$x_p = \sqrt{\frac{2\epsilon_H N_D V_{bi}}{eN_A(N_A + N_D)}}$$ (1.53)

$$d_s = x_n + x_p = \sqrt{\frac{2\epsilon_H(N_A + N_D)V_{bi}}{eN_A N_D}}$$ (1.54a)

$$E_{max} = \sqrt{\frac{2eN_A N_D}{\epsilon_H(N_A + N_D)}} V_{bi}$$ (1.55a)

where V_{bi} is the built-in voltage or built-in potential.

If we supply an external voltage $V > 0$ the inner potential difference is decreased: $(V_{bi} - V) < V_{bi}$ as shown in Fig. 1.28(f). On the other hand, if $V < 0$ the potential difference across the p-n junction is increased (Fig. 1.28(c)). This causes a change in the space-charge layer width (Lion *et al.*. 1987)

$$d_s = \sqrt{\frac{2\epsilon_H(N_A + N_D)}{eN_A N_D}} (V_{bi} - V)$$ (1.54b)

and the maximum field

$$E_{max} = \sqrt{\frac{2eN_A N_D}{\epsilon_H(N_A + N_D)}} (V_{bi} - V)$$ (1.55b)

The dashed lines in Fig. 1.28 are valid for $V = 0$.

Normally we have one-sided (unsymmetrically doped) p-n junctions with $N_A \gg N_D$ (Fig. 1.27(b)). In this case we have

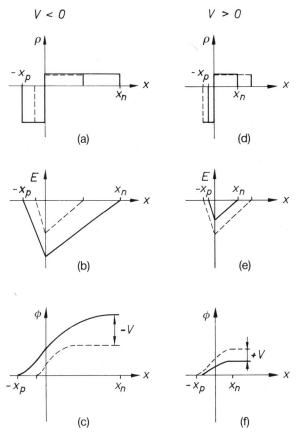

Fig. 1.28. Space charge, electric field, and potential at the p-n junction for: (a)–(c) negative and (d)–(f) positive biases.

$$d_s \approx x_n = \sqrt{\frac{2\,\epsilon_H}{eN_D}\,(V_{bi} - V)} \qquad (1.54c)$$

and

$$E_{max} = \sqrt{\frac{2eN_D}{\epsilon_H}\,(V_{bi} - V)} \qquad (1.55c)$$

If we supply a negative voltage $V < 0$ which is high enough to create a space-charge layer that penetrates the entire low-doped n-side ($x_n = d_n$) we see an effect called punch-through which causes a rapid increase of the reverse current. Punch-through can happen only if the maximum field, E_{max}, in the space-charge layer at punch-through is lower than the critical field, E_{crit}, for avalanche breakdown. This is true if the doping concentration, N_D, of the low-doped n-side is

$$N_D \leq \frac{\epsilon_H E_{crit}}{e \, d_n}. \qquad (1.56)$$

For higher impurity concentrations the high electric field causes impact ionization in the space-charge layer and avalanche breakdown. For silicon, the critical field is approximately $E_{crit} = 5 \times 10^5$ V/cm.

The avalanche breakdown voltage can be calculated from the avalanche integral, which is given by (Sze 1969)

$$\int_{-x_p}^{x_n} \alpha(E) \, dx = 1, \qquad (1.57)$$

if we assume equal ionization coefficients for electrons and holes $(\alpha_n(E) = \alpha_p(E) = \alpha(E)$; see Fig. 1.7). Using the field distributions $E(x)$ in the space-charge region $-x_p \leq x \leq x_n$ (see, for example, Fig. 1.28(b)) we derive, by numerical calculations, results for breakdown voltages of p-n junctions as shown in Fig. 1.29. Because of the curvature of a real p-n junction there are field peaks at the edges resulting in a lower breakdown voltage V_{BR}. This is also depicted in Fig. 1.29. The smaller the depth, x_B, of the p-n junction, the greater the curvature at the edges and the lower the breakdown voltage. As $x_B \to \infty$ we have a planar p-n junction.

Figs. 1.29 and 1.30 show how the breakdown voltage increases with the specific resistance $\rho = 1/e \, \mu_n N_D$. The specific resistance increases as the impurity concentration of the low-doped side, N_D, decreases. However, for very high specific resistances (or very low N_D) we see from Equation (1.56)

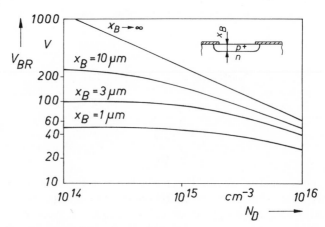

Fig. 1.29. Breakdown voltage of p-n junctions vs. impurity concentration for several junction depths x_B.

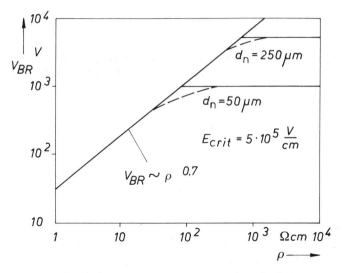

Fig. 1.30. Breakdown voltage of a p^+-n-n^+ diode.

that the punch-through occurs before avalanche breakdown. Therefore V_{BR} does not increase further at high specific resistances, as shown in Fig. 1.30. In this case the maximum field in the fully depleted low-doped n-region of width d_n is

$$E_{max} = \frac{2(V_{bi} - V)}{d_n} \qquad (1.58)$$

and at $E_{max} = E_{crit}$ the breakdown voltage $V_{BR} = V_{bi} - V \approx -V$ is given by

$$V_{BR} = \tfrac{1}{2} E_{crit} d_n. \qquad (1.59)$$

This is independent of ρ or N_D.

1.4.3 Space-charge layer capacitance

Because of the voltage dependence of the space-charge layer width (see Section 1.4.2.) we obtain a voltage-dependent space charge $Q_s(V) = eN_A x_p(V) = eN_D x_n(V)$ (Möschwitzer and Lunze 1990). This leads to a capacitance called the space-charge layer capacitance

$$C_s = \frac{dQ_s}{dV}. \qquad (1.60)$$

For a planar p-n junction we get a space-charge layer capacitance per unit area

$$C_s'' = \frac{\epsilon_H}{d_s(V)} \qquad (1.61)$$

where ϵ_H is the permittivity of the semiconductor. For an abrupt p-n junction with space-charge layer width $d_s(V)$ we find from Equation (1.54b)

$$C_s'' = \sqrt{\frac{e\,\epsilon_H N_A N_D}{2(N_A + N_D) V_{bi}}} \left(1 - \frac{V}{V_{bi}}\right)^{-\frac{1}{2}} \qquad (1.62)$$

or

$$C_s'' = C_{so}'' \left(1 - \frac{V}{V_{bi}}\right)^{-\frac{1}{2}} \qquad (1.63)$$

where V_{bi} is the built-in voltage.

In Fig. 1.31 we have plotted $1/C_s''^2$ versus V. This plot is a straight line with the intersection of the abscissa at $V = V_{bi}$. From the slope of the line we can estimate the doping concentration. For Gaussian impurity profiles (Equation (1.42)) the space-charge layer capacitance can be calculated by solving Poisson's equation numerically. Some of these results are given in Fig. 1.32. The effect of the voltage dependence of the space-charge layer capacitance is used in space-charge varactor diodes (Yang 1978).

1.4.4 Current–voltage characteristics

If we supply a positive voltage V, as shown in Fig. 1.33 a forward current I flows across the p-n junction. This current is composed of 3 parts:

(1) a hole injection current into the n-side $I_p(x_n)$;

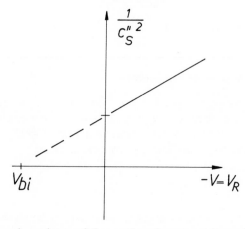

Fig. 1.31. Voltage dependence of the p-n junction space-charge layer capacitance.

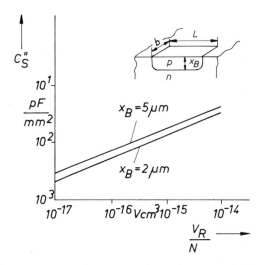

Fig. 1.32. Voltage dependence of the space-charge layer capacitance of diffused p-n junctions.

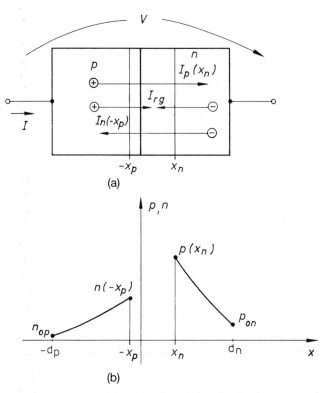

(a)

(b)

Fig. 1.33. (a) Currents and (b) minority carrier distributions at p-n junctions.

(2) an electron injection current into the p-side $I_n(-x_p)$;

(3) a recombination–generation current in the space-charge layer I_{rg}.

First we consider the recombination–generation current in the space-charge layer by solving the continuity equation (1.19) for stationary conditions ($\partial p/\partial t = 0$) and $R - G = R(n, p)$ of Equation (1.10b). In one dimension we have

$$\frac{dI_p}{dx} = -eAR(n, p) \tag{1.64}$$

and

$$I_{rg} = I_p(-x_p) - I_p(x_n) = eA \int_{-x_p}^{x_n} R(n, p)\, dx. \tag{1.65}$$

The result of a non-trivial analysis of Equation (1.65) is given here without proof (Möschwitzer and Lunze 1990; Shockley and Read 1952)

$$I_{rg} = eA \frac{n_i}{\tau_s} d_s(V) \frac{\exp\left(-\dfrac{V}{2V_T}\right)}{1 + \exp\left(-\dfrac{V}{2V_T}\right)} \left(\exp\frac{V}{V_T} - 1\right) \tag{1.66}$$

where τ_s is a carrier lifetime in the space-charge layer. For positive bias ($V \gtrsim 0.5$ V), I_{rg} is very low and can be neglected. For negative bias we obtain, from Equation (1.66), the approximation

$$I_{rg} = -eA \frac{n_i}{\tau_s} d_s(V) = I_R. \tag{1.67}$$

The injection currents, $I_p(x_n)$ and $I_n(-x_p)$ can be derived from the charge carrier distributions in the neutral regions (see Fig. 1.33) by solving the continuity and current-flow equations (1.16)–(1.19). At the space-charge layer boundaries we have the following increased minority carrier concentrations (Möschwitzer and Lunze 1990; Muller and Kamins 1977; Shockley 1950; Sze 1969; Yang 1978)

$$p(x_n) = p_{on} \exp\frac{V}{V_T} \tag{1.68a}$$

$$n(-x_p) = n_{op} \exp\frac{V}{V_T} \tag{1.68b}$$

(see also Fig. 1.33(b)).

Let us now consider the hole injection current into the n-side. (The electron injection current into the p-side is treated similarly.) The hole current equation under d.c. (stationary) conditions is, in one-dimensional form,

$$I_p = eAD_p \left(\frac{E_n}{V_T} p - \frac{dp}{dx} \right) \tag{1.69}$$

where we have used $\mu_p = D_p/V_T$ and $V_T = kT/e$. The field E_n in the n-region can be estimated from the electron (majority) carrier current equation in the n-side:

$$I_n = eAD_n \left(\frac{E_n}{V_T} n + \frac{dn}{dx} \right). \tag{1.70}$$

The majority carrier concentration n cannot be disturbed significantly from its equilibrium value $n \approx n_{on} \approx N_D$ at reasonable (low) currents. Therefore we set $n = N_D$ in Equation (1.70). In any case I_n is very low in comparison to the drift and diffusion parts. Therefore in Equation (1.70) we set $I_n \approx 0$ and the drift field in the n-region is

$$E_n = - \frac{V_T}{N_D} \frac{dN_D}{dx}. \tag{1.71}$$

For homogeneously doped semiconductors ($N_D = $ const.) we have no drift field ($E = 0$).

If we insert E_n of Equation (1.71) into Equation (1.69) we get

$$I_p = - eAD_p \left(\frac{p}{N_D} \frac{dN_D}{dx} + \frac{dp}{dx} \right) \tag{1.72a}$$

or

$$I_p N_D \, dx = - eA D_p \, d(N_D p). \tag{1.72b}$$

The integration of Equation (1.72b) yields

$$I_p = eA D_p \frac{\left((N_D p)_{x_n} - (N_D p)_{d_n} \right)}{\int_{x_n}^{d_n} N_D \, dx}. \tag{1.73}$$

With the boundary conditions

$$(N_D p)_{x_n} = n_{on} p_{on} \exp \frac{V}{V_T} = n_i^2 \exp \frac{V}{V_T} \tag{1.74}$$

and

$$(N_D p)_{d_n} = n_{on} p_{on} = n_i^2 \tag{1.75}$$

we obtain the hole injection current

$$I_p = eA D_p \frac{n_i^2}{Z_D} \left(\exp \frac{V}{V_T} - 1 \right) \tag{1.76}$$

where Z_D is the density, per unit area, of donor atoms in the n-region (the Gummel number)

$$Z_D = \int_{x_n}^{d_n} N_D \, dx. \tag{1.77}$$

If we repeat all these calculations for electrons in the p-region we get, for the electron injection current into the p-region,

$$I_n = eAD_n \frac{n_i^2}{Z_A} \left(\exp \frac{V}{V_T} - 1 \right) \tag{1.78}$$

where the content, per unit area, of acceptors in the p-region is

$$Z_A = \int_{-d_p}^{-x_p} N_A \, dx. \tag{1.79}$$

The complete I/V characteristic of a p-n junction, with I_{rg} of Equation (1.66), I_p of Equation (1.76), and I_n of Equation (1.78), is

$$I = eAn_i^2 \left\{ \frac{D_n}{Z_A} + \frac{D_p}{Z_D} + \frac{d_s}{n_i \tau_s} \frac{\exp\left(-\dfrac{V}{2V_T}\right)}{1 + \exp\left(-\dfrac{V}{2V_T}\right)} \right\} \left(\exp \frac{V}{V_T} - 1 \right). \tag{1.80}$$

This I/V characteristic is depicted in Fig. 1.34.

In the reverse direction (that is, for negative bias $V < -0.5$ V) I_{rg} dominates in Si and GaAs diodes (because of the small intrinsic density n_i) and we have, for the reverse current,

$$-I = I_R = eA \frac{n_i d_s}{\tau_s}. \tag{1.81a}$$

On the other hand, for Ge diodes (large n_i), we get

$$-I = I_R = eAn_i^2 \left(\frac{D_n}{Z_A} + \frac{D_p}{Z_D} \right). \tag{1.81b}$$

For large negative voltages a rapid increase of the current occurs at the breakdown voltage $-V = V_{BR}$ because of avalanche and/or punch-through effects. For small positive voltages ($V < 0.5$ V) we get, for Si and GaAs diodes,

$$I_F = eA \frac{n_i d_s}{\tau_s} \exp \frac{V}{2V_T}, \tag{1.82a}$$

and, for medium and large positive voltages, ($V > 0.5$ V),

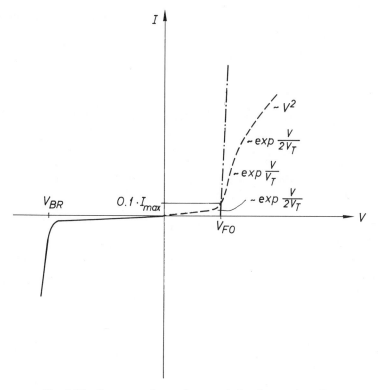

Fig. 1.34. Current–voltage characteristic of a p-n junction.

$$I_F = eAn_i^2 \left(\frac{D_n}{Z_A} + \frac{D_p}{Z_D} \right) \exp \frac{V}{V_T}. \qquad (1.82b)$$

For very high currents, high-injection effects, such as conductivity modulation, voltage drops, and space-charge limitation, result in the current depending on $\exp(V/2V_T)$ and, after that, on V^2 as sketched in Fig. 1.34 with dashed lines (Möschwitzer and Lunze 1990; Muller and Kamins 1977; Shockley 1950; Sze 1969; Yang 1978). To include some of these effects the I/V characteristic approximation

$$I_F = I_S \left(\exp \frac{V}{mV_T} - 1 \right) \qquad (1.83)$$

with $m \geq 1$ can be used.

To describe the behaviour of the forward current flow we define a diode forward floating potential, V_{FO}, where the current is 10% of its maximum value (see Fig. 1.34). V_{FO} does not depend much on the current. It is, for silicon diodes $V_{FO} \approx 0.7\,\text{V}$, and, for germanium diodes, $V_{FO} \approx 0.4\,\text{V}$.

Because of the exponential rise of the current we can consider the diode as a switch that is on for $V \geq V_{FO}$ and off for $V < V_{FO}$. This switching characteristic is also shown in Fig. 1.34 (dotted line).

1.4.5 Heterojunctions

Junctions between two different semiconductors are called heterojunctions (e.g. GaAlAs–GaAs (Möschwitzer and Lunze 1990; Muller and Kamins 1977; Sze 1969; Yang 1978)). There are many possible electric effects which can be used successfully to construct semiconductor and optoelectronic devices (Milnes 1986). From an engineering point of view such heterojunctions are very important. For these we have, at the interface, only a very small distortion (discontinuity) of the crystal structure so that most of the electronic behaviour is like that for a homojunction. Therefore the analytic expressions derived in Sections 1.4.1. and 1.4.4. can be applied (approximately) to heterojunctions too. The minority carrier injection currents of a hetero- p-n junction are

$$I_p(x_n) = eAD_p \frac{n_{in}^2}{Z_D} \left(\exp \frac{V}{V_T} - 1\right) \qquad (1.84)$$

$$I_n(-x_p) = eAD_n \frac{n_{ip}^2}{Z_A} \left(\exp \frac{V}{V_T} - 1\right). \qquad (1.85)$$

Their ratio is

$$\Gamma = \frac{D_p n_{in}^2 Z_A}{D_n n_{ip}^2 Z_D}. \qquad (1.86)$$

Because of the different band gaps and intrinsic densities n_{in} and n_{ip} we can produce high injection ratios Γ at equal impurity contents Z_D and Z_A by choosing proper materials with $n_{ip} \gg n_{in}$ (or vice versa). This is important for heteroemitter transistors.

1.4.6 Semiconductor diodes as rectifiers

Because of the non-linear I/V characteristic of a semiconductor diode shown in Fig. 1.34 it can be applied as a rectifier of a.c. currents. A basic circuit for doing this is shown in Fig. 1.35(a). The a.c. voltage, $V_{\sim}$, is supplied by a transformer and is connected via a diode to a load resistor, R. Only during the positive half wave of the a.c. voltage is a current flow possible and we have on R a pulsing d.c. voltage (see Fig. 1.35(c)). If we connect a capacitor, C, in parallel to R the d.c. voltage is smoothed depending on the ratio of the time constant RC and the period of the a.c. voltage. The capacitor C is charged during the current flow angle, Θ, where V is greater than $V_0 + V_{FO}$

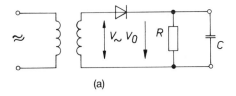

(a)

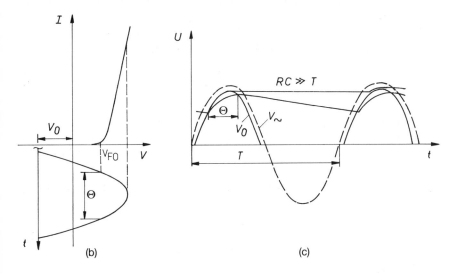

(b)   (c)

Fig. 1.35. Rectification with p-n junction diodes.

(Fig. 1.35(b)). If $V < V_0 + V_{F0}$ the capacitor C is discharged (see Fig. 1.35(c)).

If $RC \gg T$ the capacitor will be only slightly discharged and the d.c. voltage is nearly constant:

$$V_0 = \hat{V}_- - V_{F0}. \qquad (1.87)$$

This is applied in radio engineering and is called AM demodulation. For the demodulation (rectification) of small a.c. signals we need diodes with small floating potentials V_{F0} (e.g. germanium, Schottky, and backward diodes (Möschwitzer and Lunze 1990)).

For rectification of large voltages (power rectifiers) we need diodes with high breakdown voltages, V_{BR} (see Section 1.4.2). High breakdown voltages ($V_{BR} > 50\,\text{V}$) cannot be realized with pure p-n junctions. Therefore for power rectifiers we use p-i-n, p^+-n-n^+, or p^+-p-n^+ diodes. At reverse bias the intrinsic (i) or weakly doped (n or p) layer between the heavily doped p^+ and n^+ layers is fully depleted. The field in the intrinsic layer (width d_i) is

$$E = \frac{V}{d_1}, \tag{1.88}$$

and the breakdown voltage

$$V_{BR} = E_{crit} d_1 \tag{1.89}$$

can be adjusted by designing the width d_1 of the intrinsic layer (e.g. for $V_{BR} = 1000\,V$, $E_{crit} = 5 \times 10^5\,V\,cm^{-1}$ we need $d_1 = 20\,\mu m$).

The maximum field in a fully depleted n-type layer (width d, doping concentration N) is, by solving Poisson's equation,

$$E_{max} = \frac{eNd}{\epsilon_H} + E_{min} \tag{1.90}$$

and the field distribution

$$E(x) = \frac{eNx}{\epsilon_H} + E_{min}. \tag{1.91}$$

Neglecting all voltage drops except that on the depleted n-region we obtain, by integrating Equation (1.91), the voltage

$$V = \frac{eNd^2}{2\epsilon_H} + E_{min} d \tag{1.92}$$

and we get

$$E_{min} = \frac{V - \dfrac{eNd^2}{2\epsilon_H}}{d}.$$

If we insert E_{min} into Equation (1.90) we get maximum field

$$E_{max} = \frac{V}{d} + \frac{eNd}{2\epsilon_H} \tag{1.93}$$

and finally the breakdown voltage for this p^+-n-n^+ rectifier diode

$$V_{BR} = E_{crit} d - \frac{eNd^2}{2\epsilon_H}. \tag{1.94}$$

This is smaller than for the ideal p-i-n diode. Therefore the goal is to make the doping of the n-layer as small as possible.

If the i- or n-layer is made too thick ($d > 100\,\mu m$) the forward current characteristic will be degraded (Möschwitzer and Lunze 1990). At forward biases a large number of electrons and holes are injected into the low-doped middle layer. This makes this layer highly conductive. The injected charge carrier densities are

$$n = p = n_i \exp \frac{V}{2V_T}. \tag{1.95}$$

These are nearly constant throughout the middle layer if this layer is not too thick. If we assume an electron–hole lifetime of τ we obtain the I/V characteristic of an ideal p-i-n diode

$$I = eA \frac{n_i d}{\tau} \left(\exp \frac{V}{2V_T} - 1 \right) \tag{1.96}$$

which is mainly due to the recombination current in the i- or low-doped middle layer.

1.4.7 Semiconductor switching and charge-storage varactor diodes

For digital circuits the p-n diode can be applied as an electronic switch because of its switching characteristic shown in Fig. 1.34. Two basic switching circuits with p-n diodes are sketched in Fig. 1.36. In the circuit of Fig. 1.36(a) a forward current, I_F, is supplied to the diode when the pulse voltage, V_i, is zero. From Fig. 1.36(a) we derive

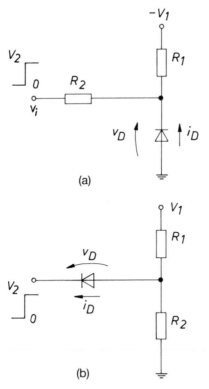

Fig. 1.36. Switching circuits with p-n junction diodes.

$$I_F = \frac{V_1}{R_1} - V_{FO}\left(\frac{1}{R_1} + \frac{1}{R_2}\right) \approx \frac{V_1}{R_1}. \tag{1.97}$$

If we switch V_i from 0 to V_2 a reverse current is supplied

$$I_R = -\frac{V_2 + V_{FO}}{R_2} + \frac{V_1 - V_{FO}}{R_1} \approx -\frac{V_2 R_1 - V_1 R_2}{R_2 R_1}. \tag{1.98}$$

For this reason the diode voltage drop cannot change immediately and therefore stays positive ($V_D \approx V_{FO}$) for a moment until the diode is discharged (see Fig. 1.37(a)).

The time response of the diode current after a switching operation at $t = 0$ from forward (I_F) to reverse ($-I_R$) current is shown in Fig. 1.37(b). During the storage time t_s the reverse current $-I_R$ is nearly constant and the diode will be discharged. The diode voltage (see Fig. 1.37(a)) changes from

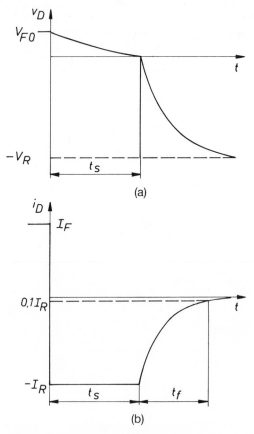

(a)

(b)

Fig. 1.37. Switching response of (a) diode voltage and (b) diode current.

$V_D = V_{FO}$ at the beginning to $V_D = 0$ at the end of the storage phase. For $t \geq t_s$ the diode is nearly discharged and the current decreases rapidly. At the end of this recovery phase the reverse voltage at the diode is

$$V_R = -\frac{R_1 V_2 - R_2 V_1}{R_1 + R_2}. \qquad (1.99)$$

This behaviour can be explained by the charge dynamics in the diode as shown in Fig. 1.38. At the beginning ($t = 0$) we have a steady state charge carrier distribution determined by the forward current. The boundary concentrations for electrons and holes are

$$p(x_n) = p_{on} \exp\frac{V_{FO}}{V_T} \qquad (1.100)$$

and

$$n(-x_p) = n_{op} \exp\frac{V_{FO}}{V_T} \qquad (1.101)$$

respectively (see Fig. 1.38).

At the moment of current switching ($t = 0$) from $i_D = +I_F$ to $i_D = -I_R$ (see Fig. 1.37(b)) only the gradients of the charge carrier distributions, dn/dx and dp/dx, are altered. A finite change in the total charge is

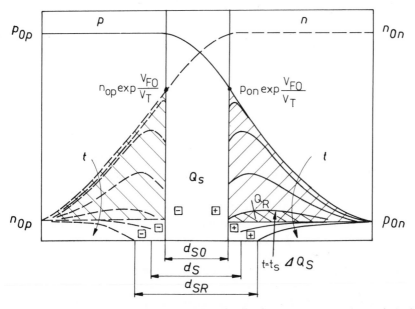

Fig. 1.38. Response of the minority carrier distribution at a p-n junction switched from forward to reverse bias (reverse recovery).

impossible at $t = 0$ (because $i = dQ/dt$). From now on the discharges $p(x, t)$ and $n(x, t)$ proceed until at $t = t_s$, $p(x_n, t_s) = p_{on}$ and $n(-x_p, t_s) = n_{op}$ and therefore $\cdot V_D = 0$. We now have only a very small residual charge Q_R in the diode and the current decreases rapidly as mentioned above. The analysis of this dynamic switching process can be carried out by using the charge-control equation

$$i_D = \frac{dQ_B}{dt} + \frac{Q_B}{\tau_V} \qquad (1.102)$$

where Q_B is the total charge (electrons and holes) stored in p- and n-regions of the diode, and τ_V is a parameter depending on the lifetime τ and the transit time τ_B of the charge carriers $(1/\tau_V = 1/\tau + 1/\tau_B)$.

For steady state forward current flow we have $i_D = I_F$ and $dQ_B/dt = 0$ and we have a charge

$$Q_{BS} = I_F \tau_V \qquad (1.103)$$

stored in the diode. During the reverse current flow $i_D = -I_R$ we obtain

$$-I_R = \frac{dQ_B}{dt} + \frac{Q_B}{\tau_V} \qquad (1.104)$$

and, with the initial condition $Q_B(0) = Q_{BS}$, we derive from Equation (1.104) the time response of the charge stored in the diode during the discharging phase

$$Q_B(t) = \tau_V (I_R + I_F) \exp\left(-\frac{t}{\tau_V}\right) - \tau_V I_R. \qquad (1.105)$$

If we neglect any residual charge at $t = t_s$ we can estimate, from Equation (1.105) with $Q_B(t_s) = 0$, the storage time

$$t_s = \tau_V \ln\left(1 + \frac{I_F}{I_r}\right) \qquad (1.106)$$

For $t > t_s$ we have the fall time t_f (recovery phase). t_f is mainly determined by the discharge of the residual charge and the space-charge layer capacitor. Diodes that are specially designed for very short fall times are called *step recovery* or *charge-storage* varactor diodes (Schaffner 1970). They can be applied as very fast switching elements in sampling equipment or as upward converters for microwave applications.

In Fig. 1.39 we show the current of a charge-storage varactor diode which is driven by a sinusoidal voltage. Because of the step recovery (very short fall time) the reverse current decays in a very short time period. This allows the generation of higher harmonics in the microwave range. In Fig. 1.40 the amplitudes of three harmonics are shown as a function of the stepping angle for a typical charge-storage varactor diode.

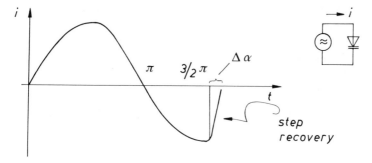

Fig. 1.39 Current waveform at a charge-storage varactor diode.

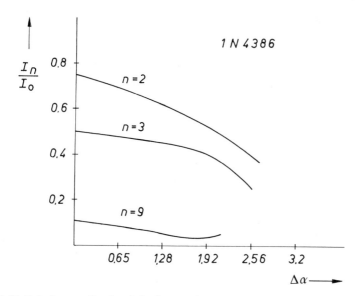

Fig. 1.40 Relative amplitude of the harmonics at a charge-storage varactor diode.

1.4.8 Temperature dependence

The forward current I_F is proportional to n_i^2, by Equation (1.82), and n_i^2 is strongly temperature dependent $(n_i^2 \sim \exp(-Wg/kT))$ as in Equation (1.7). This yields an exponential increase of the forward current. On the other hand, the voltage factor $\exp(V/V_T) = \exp(eV/kT)$ causes a decrease of the diode forward current with increasing temperature. In summary the exponential temperature dependence of the forward current can be written as

$$I_F = I_F(T_o) \exp\{c_f(T - T_o)\} \qquad (1.107)$$

where T_o is a reference temperature (e.g. 300 K) and c_f is the exponential temperature coefficient

$$c_f = \frac{W_g - eV}{kT_o^2} \tag{1.108}$$

where W_g is the energy gap.

The reverse current, I_R, is also exponentially dependent on the temperature

$$I_R(T) = I_R(T_o) \exp\{c_r(T - T_o)\} \tag{1.109}$$

where

$$c_r = \frac{W_g}{mkT_o^2} \tag{1.110}$$

($m = 1$ for germanium, $m \approx 2$ for silicon and GaAs).

1.5   METAL–SEMICONDUCTOR CONTACTS

1.5.1   Overview

There are various types of I/V characteristics of a metal–semiconductor contact (junction) depending on the metal, the contact preparation (interface properties), and the doping of the semiconductor (Card 1976). Some of these are sketched in Fig. 1.41. Whereas on p-type silicon, aluminium forms a low-resistive ohmic contact, on low-doped n-type silicon it normally forms a Schottky diode having an I/V characteristic

$$I = I_S(\exp\frac{V}{V_T} - 1). \tag{1.111}$$

On heavily doped semiconductors a metal contact normally gives low resistive ohmic contacts. (We also get very good ohmic contacts with silicides of refractory metals such as WSi_2, $TiSi_2$ $MoSi_2$).

For special applications multilayer contact structures (e.g. AlSiTi, AlSiCu) can be applied to achieve good contacts with low electromigration or to form diffusion barriers (e.g. AlTiW) for contacts on very shallow semiconductor layers. In Table 1.2 some parameters of contact systems are summarized.

The I/V characteristics (a) and (b) in Fig. 1.41 can be explained by means of the energy-band diagrams in Fig. 1.42. There is a potential barrier V_B at the metal–semiconductor interface. For a metal– n-silicon contact, V_{Bn} gives a depletion region x_s at the surface. This means the electron concentration decreases towards the interface. Because of this decrease of majority carriers we have a high resistive sheet at the interface. This is

contacts

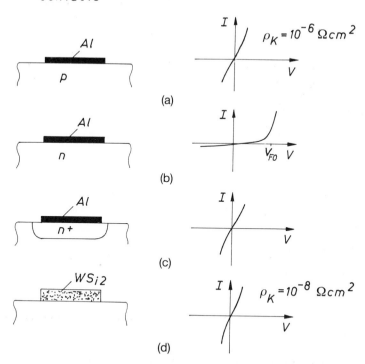

Fig. 1.41. Characteristics of different types of metal–semiconductor contacts.

Table 1.2. Contact system.

contact material	$\rho_K(\Omega\, cm^2)$	$R_K = \dfrac{\rho_K}{A_K}\,(\Omega)$
Al	10^{-6}	< 10
Al (TiW)		for contacts
AlSiTi	2.10^{-7}–10^{-6}	on silicon
ALSiCu		
W-WSi$_2$	10^{-7}–10^{-6}	< 1
Mo-MoSi$_2$		for contacts
MoSi$_2$	10^{-7}–10^{-6}	of interconnections

equivalent to the space-charge layer of a p-n junction. This depletion region increases for negative biases (Fig. 1.42(a$_3$)) and prevents current flow. For positive biases (Fig. 1.42(a$_2$)) electrons are injected into the space-charge layer and into the metal and a forward current like that of a p-n junction is possible (see Section 1.5.2).

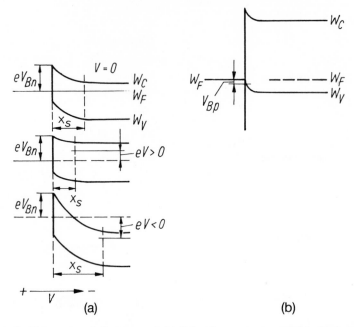

Fig. 1.42. Energy-band diagrams of (a) Schottky contacts and (b) ohmic contact.

For a metal–p-silicon contact (Fig. 1.42(b)) we have a barrier V_{Bp} for holes which gives an accumulation layer at the interface and the hole concentration increases towards the interface. This accumulation layer at the interface has a low resistivity which yields an ohmic contact (see Fig. 1.41(a)). This contact is needed for the interconnection of functional elements (devices) in integrated circuits.

Let us now consider the contact of a metal on a very heavily doped semiconductor (Fig. 1.41(c)). Although we have, in this case, a depletion layer, (Fig. 1.43), we get a low resistive contact because of a tunnelling

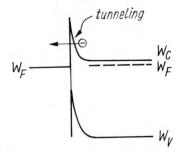

Fig. 1.43 Energy-band diagram of a metal–semiconductor contact with tunnelling.

current which flows through the very thin potential barrier, as shown in Fig. 1.43. From Section 1.4.2. we know that the width of a depletion or space-charge layer decreases with increasing impurity concentration N (see Equation (1.54c)), so the depletion layer width in a heavily doped semiconductor is very thin. In the case of tunnelling the carriers move through the potential barrier rather than over it which results in a large current and a low resistive contact.

1.5.2 Schottky diodes

As already stated, a Schottky diode is a metal–semiconductor contact with a depletion layer at the interface. The current in this diode is determined mainly by thermal emission of electrons from the semiconductor into the metal across a potential barrier ΔV_B (Möschwitzer and Lunze 1990)

$$I = eA \frac{n v_{th}}{4} \exp(-\Delta V_B) \qquad (1.112)$$

where n is the electron density in the semiconductor bulk and v_{th} is the thermal velocity (Shockley 1950)

$$v_{th} = \sqrt{\frac{8kT}{\pi m}} . \qquad (1.113)$$

The net current I is the difference between the electron emission current into the metal, I_{HM}, and the emission current from the metal into the semiconductor, I_{MH},

$$I = I_{HM} - I_{MH}. \qquad (1.114)$$

From Fig. 1.42(a) we obtain the barrier height for electrons in the semiconductor at a positive bias V

$$\Delta V_B = V_{Bn} - V - \frac{1}{e}(W_C - W_F) = V_{bi} - V \qquad (1.115)$$

and therefore the emission current into the metal

$$I_{HM} = eA n \frac{v_{th}}{4} \exp\left(\frac{-eV_{Bn}}{kT}\right) \exp\left(\frac{V}{V_T}\right) \exp\left(\frac{W_C - W_F}{kT}\right). \qquad (1.116)$$

If we use Equation (1.5a) $n = N_C \exp\{-(W_C - W_F)/kT\}$ we obtain

$$I_{HM} = eA N_C \frac{v_{th}}{4} \exp\left(-\frac{V_{Bn}}{V_T}\right) \exp\left(\frac{V}{V_T}\right). \qquad (1.117)$$

Because we have $I = 0$ for $V = 0$ we find

$$I = eA \frac{N_C v_{th}}{4} \exp\left(-\frac{V_{Bn}}{V_T}\right)\left(\exp\frac{V}{V_T} - 1\right) \qquad (1.118)$$

$$\underbrace{\phantom{I = eA \frac{N_C v_{th}}{4} \exp\left(-\frac{V_{Bn}}{V_T}\right)}}_{I_S}$$

where

$$I_S = eA \frac{N_C v_{th}}{4} \exp\left(\frac{-V_{Bn}}{V_T}\right) = eA 4\pi m k^2 T^2 \exp\left(\frac{-V_{Bn}}{V_T}\right). \qquad (1.119)$$

The I/V characteristic of a Schottky diode is like that of a p-n junction (Equation (1.83)).

The depletion layer width x_s (see Fig. 1.42(a)) depends on the supplied voltage V. As for a one-sided p-n junction (see Section 1.4.2) we get

$$x_s = \sqrt{\frac{2\epsilon_H}{eN}(V_{bi} - V)} \qquad (1.120)$$

where V_{bi} is the built-in potential at thermal equilibrium. From Fig. 1.42(a) we derive

$$eV_{bi} = eV_{Bn} - (W_C - W_F) \qquad (1.121)$$

and with $n_o = N$ and $V_T = kT/e$ we get

$$W_C - W_v = kT \ln \frac{N_C}{N} = eV_T \ln \frac{N_C}{N}. \qquad (1.122)$$

Using the voltage-dependent space-charge layer (depletion layer) width $x_s(V)$ we obtain for the space-charge layer capacitance, per unit area, of the Schottky diode

$$C_s'' = \frac{\epsilon_H}{x_s} = \sqrt{\frac{\epsilon_H eN}{2}}\left(V_{Bn} - V_T \ln \frac{N_C}{N} - V\right)^{-\frac{1}{2}} \qquad (1.123)$$

or

$$\frac{1}{C_s''^2} = \frac{2}{eN\epsilon_H}\left(V_{Bn} - V_T \ln \frac{N_C}{N} - V\right). \qquad (1.124)$$

$1/C_s''^2$ gives a straight line (see Fig. 1.44) and an intersection with the abscissa at $V = V_{Bn} - V_T \ln N_C/N$ and a slope of $(e\epsilon_H N/2)^{-1}$. These parameters can be measured easily and we can calculate the barrier height, V_{Bn}, and the doping concentration, N, of the semiconductor from the measured data.

Schottky diodes have some advantages over p-n diodes, the most important of which is the dynamic behaviour. In Schottky diodes electrons are injected into the metal (rather than as minority carriers into a p-type semiconductor), but in a metal layer they cannot be stored as an excess

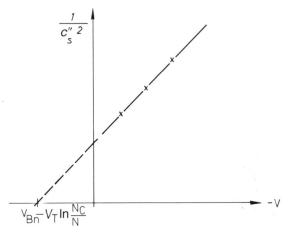

Fig. 1.44. Voltage dependence of the metal–semiconductor space-charge layer capacitance.

charge. Therefore we do not have a storage time t_s in the reverse recovery phase of the diode as we had for p-n diodes (see Fig. 1.37). Therefore Schottky diodes run much faster than p-n diodes.

1.5.3 The potential barrier of a metal–semiconductor contact

As we have seen so far the properties of metal–semiconductor contacts are determined by the potential barrier, V_B, at the interface. We will now investigate some details of this potential barrier using the model of Fig. 1.45. In a real metal–semiconductor contact we always have a very thin insulation (oxide) layer in between the metal and semiconductor of thickness δ_i. The energy-band diagram of this MOS structure is sketched in Fig. 1.45. Right at the surface of the semiconductor we have energy states (surface states with

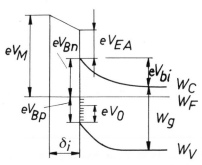

Fig. 1.45. Energy-band diagram of a real metal–semiconductor contact with a thin interface oxide layer.

the density N_{ss}' per cm^2 per eV). These energy states can be empty (above the Fermi level) or occupied (underneath the Fermi level), and they can be charged or neutral. We assume an energy level eV_o which fits to the Fermi level if the net interface charge is zero. $W_M = eV_M$ is the metal work function. Now we have to consider the following charges at that structure:

(1) Charge of the interface states, per unit area

$$Q_Z'' = -eN_{ss}'(W_g - eV_{Bn} - eV_o); \tag{1.125}$$

(2) Charge of the ionized impurities in the surface space-charge layer (depletion layer)

$$Q_B'' = eNx_s = \sqrt{2\epsilon_H N(eV_{Bn} - (W_C - W_F))}; \tag{1.126}$$

(3) Charge per unit area on the metal electrode

$$Q_T'' = \epsilon_i \frac{V_i}{\delta_1} = -\frac{\epsilon_i}{\delta_i}(V_M - V_{EA} - V_{Bn}). \tag{1.127}$$

The neutrality condition

$$Q_T'' + Q_B'' + Q_Z'' = 0 \tag{1.128}$$

yields, with the approximation

$$2\epsilon_H N(eV_{Bn} - (W_C - W_F)) \ll e^2 N_{SS}'\left(V_{Bn} + V_o - \frac{W_g}{e}\right) \tag{1.129}$$

the barrier height

$$V_{Bn} = \gamma(V_M - V_{EA}) + (1 - \gamma)\left(\frac{W_g}{e} - V_o\right) \tag{1.130}$$

where

$$\gamma = \frac{1}{1 + \dfrac{e^2 N_{ss}'\delta_i}{\epsilon_i}}. \tag{1.131}$$

For $N_{ss}' = 0$ we get $\gamma = 1$ and

$$V_{Bn} = V_M - V_{EA}, \tag{1.132}$$

which is the classical Schottky contact formula. As $N_{ss}' \to \infty$ we get

$$V_{Bn} = \frac{W_g}{e} - V_o. \tag{1.133}$$

This means the Fermi level is 'clamped' and the barrier height does not depend on the metal. In Table 1.3 some parameters for metal–semiconductor

Table 1.3. Parameters of metal–semiconductor contacts

	V_M/V		Si	Ge	GaAs	GaP
		V_{EA}/V	4.15	4.05	4.07	4.3
Al	4.1	V_{Bn}/V	0.6–0.7	0.5	0.8	1
Au	4.7	V_{Bn}/V	0.8	0.45	0.9	1.3
Cu	4.4	V_{Bn}/V	0.7–0.8	0.5	0.8	1.2
Pt	5.4	V_{Bn}/V	0.9		0.86	1.45

contacts are summarized. PtSi and AlSi Schottky diodes are of interest mainly for Schottky TTL (see Fig. 2.26 and Section 3.1.3).

1.6 INTERCONNECTION LINES

The interconnection of single devices in integrated circuits is accomplished by using thin metal (Al, Mo, W) and doped silicon or silicide (WSi$_2$, MoSi$_2$, TiSi$_2$) layers. All these layers have a resistance per unit length, R', and a capacitance per unit length, C' to their neighbourhoods, mainly to the semiconductor substrate (see Fig. 1.46(a)). A simple model of an

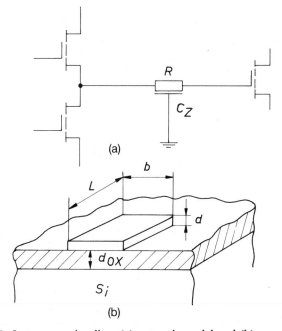

(a)

(b)

Fig. 1.46. Interconnection line: (a) network model and (b) cross-section.

interconnection line is shown in Fig. 1.46(b). A thin conductive layer, with sheet resistance R_s, length L, and width b, runs above an insulation layer (e.g. SiO_2 layer with thickness d_{ox}). The resistance per unit length is

$$R' = R_s \frac{1}{b}$$

(1.134)

and the capacitance per unit length is

$$C' = \frac{\epsilon_{ox}}{d_{ox}} b$$

(1.135)

or, for very narrow widths,

$$C' = \epsilon_{ox} \left\{ 1.15 \frac{b}{d_{ox}} + 2.8 \left(\frac{d}{d_{ox}} \right)^{0.222} \right\}$$

(1.136)

where d is the thickness of the line.

Many problems arise with these interconnection lines. Firstly the voltage drop, $\Delta V = IR$, along the line causes a change of the logic levels and therefore can yield logical errors (see Chapter 3), and secondly the signal delay may be longer than the delay caused by the individual devices. In VLSI and ULSI circuits the scaled-down devices can run faster but the interconnection delay increases because of the relatively longer interconnection lines which have higher resistivity and more parasitic capacitors. An exact analysis of this phenomenon must consider the distributed RC line with all non-linearities. For our very simple considerations we use only the RC time constant of the interconnection line as a figure of merit for the signal delay:

$$RC = R'C'L^2 = R_s \frac{\epsilon_{ox}}{d_{ox}} L^2$$

(1.137)

or, with Equation (1.136),

$$RC = R_s L^2 \frac{\epsilon_{ox}}{b} \left\{ 1.15 \left(\frac{b}{d_{ox}} \right) + 2.8 \left(\frac{d}{d_{ox}} \right)^{0.222} \right\}$$

(1.138)

L^2 is proportional to the chip area, A_C, which we can expect to increase as the technology advances; R_S can be held nearly constant if we use advanced interconnection materials; and d_{ox} and b decrease with device scaling for VLSI and ULSI. In Table 1.4 we summarize some sheet resistances for interconnection materials.

Table 1.4. Sheet resistances for interconnection lines

material	$R_S(\Omega)$
Al	0.02–0.08
W	0.15–0.4
Mo	0.15–0.4
TiSi$_2$	0.3–1
TaSi$_2$	0.7–2
WSi$_2$	1–3
MoSi$_2$	1–3
doped Poly-Si	20–100
diffused layer	10–200

1.7   PROBLEMS FOR CHAPTER 1

1. A pure wafer of silicon is doped with $N_D = 10^{16}\,\mathrm{cm}^{-3}$ phosphorus atoms and $N_A = 10^{15}\,\mathrm{cm}^{-3}$ boron atoms. The doping atoms are distributed homogeneously throughout the wafer and are fully ionized.
   (a) What is the conductivity of the silicon wafer?
   (b) Calculate the Fermi level with respect to the conduction band edge. (Use: $kT/e = 0.025\,\mathrm{V}$; effective density of states in conduction band, $N_C = 10^{19}\,\mathrm{cm}^{-3}$; electron mobility $\mu_n = 2\,\mu_p = 1000\,\mathrm{cm}^2/\mathrm{Vs}$.)

2. In a field-free semiconductor we have a hole distribution as shown in Fig. 1.47. What is the hole current density if the holes have a mobility of $\mu_p = 500\,\mathrm{cm}^2/\mathrm{Vs}$? (Use: $kT/e = 0.025\,\mathrm{V}$.)

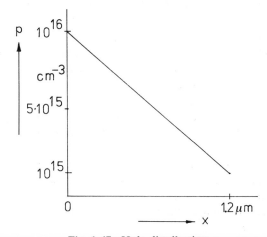

Fig. 1.47. Hole distribution.

3. Derive the electric fields in an n-type semiconductor with a donor distribution

$$N_D(x) = N_{DO} \exp(-x/L_D)$$

in thermal equilibrium.

(Use: diffusion length $L_D = 1\,\mu m$; $kT/e = 0.025\,V$.)

4. A wafer of pure silicon is illuminated with light and we have, throughout the sample, a steady state and uniform photogeneration rate of $G = 10^{20}\,cm^{-3}\,s^{-1}$.

   (a) Calculate the electron and hole densities for equal electron and hole lifetimes $\tau_n = \tau_p = 1\,\mu s$.

   (b) Determine the time, t_H, at which the excess carrier densities $\Delta n,\,\Delta p$ have decayed to half of the maximum value after the light is turned off.

   (Use: $n_i = 1.5 \times 10^{10}\,cm^{-3}$; $kT/e = 0.025\,V$; $R = \Delta n/\tau_n = \Delta p/\tau_p$.)

5. An integrated diffused resistor is fabricated in an n-epitaxy layer having a donor concentration of $N_D = 10^{16}\,cm^{-3}$. The surface concentration of the Gauss-shaped acceptor doping profile is $N_{AO} = 10^{18}\,cm^{-3}$ and the diffusion length is $L_D = 1\,\mu m$.

   (a) What length-to-width ratio L/b in the resistor layout is needed if we want to design a resistor of $R = 2\,k\Omega$ (use the diagram in Fig. 1.17).

   (b) What is the depth, x_B, of the resistor sheet (p-n junction)?

6. A p-n junction is reverse biased and has a space-charge layer width $d_s = 1\,\mu m$. The donor concentration on the n-side is $N_D = 10^{16}\,cm^{-3}$ and the acceptor concentration on the p-side is $N_A = 5 \times 10^{16}\,cm^{-3}$.

   (a) Calculate the penetration of the space-charge layer in the n-side, x_n, and the p-side, x_p, respectively.

   (b) Explain what happens for $N_A \gg N_D$ (e.g. $N_A = 10^{20}\,cm^{-3}$).

7. Under forward bias conditions holes are injected in the n-side of a p-n junction.

   (a) Calculate the hole injection current density for $V = 0.6\,V$ if the n-side is uniformly doped with $N_D = 10^{16}\,cm^{-3}$ donors and its width, d_n, is very small (recombination can be neglected).

   (Use: $n_i = 1.5 \times 10^{10}\,cm^{-3}$; $\mu_p = 500\,cm^2/Vs$; $d_n = 1\,\mu m$; $kT/e = 0.025\,V$; $A = 100\,\mu m^2$.)

   (b) Determine the hole charge stored in the n-region.

   (c) What is the time constant τ_V for charge storage (see Section 1.4.7.)?

8. A one-sided abrupt p-n junction has an acceptor concentration $N_A = 10^{19}\,cm^{-3}$ and a donor concentration $N_D = 10^{17}\,cm^{-3}$.

   (a) Determine the built-in potential V_{bi}.

   (b) Calculate the space-charge layer width and the maximum field for $V = -10V$.

   (Use: $kT/e = 0.025\,V$; permittivity of silicon $\epsilon_H = 10^{-12}\,As/Vcm$; $n_i = 1.5 \times 10^{10}\,cm^{-3}$.)

9. In Fig. 1.48 a measured capacitance–voltage characteristic is given

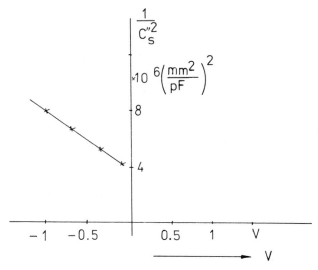

Fig. 1.48. Voltage dependence of a space-charge layer capacitance.

for a one-sided step p-n junction. Estimate the built-in potential, V_{bi}, and the donor concentration, N_D, of the lightly doped side ($\epsilon_H = 10^{-12}$ As/Vcm).

10. A one-sided step p-n junction is given with $N_A \gg N_D$.

(a) What is the donor concentration in the lightly doped n-side with a width of $d_n = 1\,\mu m$ for which punch-through occurs at the maximum field $E_m = 3 \times 10^5$ V cm^{-1}.

(b) Calculate the punch-through voltage for this case.

(Use: permittivity $\epsilon_H = 10^{-12}$ As/Vcm; built-in potential $V_{bi} = 0.8$ V.)

11. What is the a.c. small-signal resistance of a p-n junction at $I = 1$ mA d.c. current? (Use: $kT/e = 0.025$ V.)

12. A diode switching circuit is shown in Fig. 1.49. Derive the reverse recovery time t_s (storage time) for a diode which has at a forward current of $I_F = 1$ mA a stored charge $Q_{BS} = 10^{-11}$ As.

13. An input voltage, V_i, has to be stabilized at a voltage, V_o, by using a Z-diode. The load current is $I_o = 1$ A. Design a circuit, calculate the circuit parameters, and find the maximum power of the Z-diode.

14. (a) Sketch the energy-band diagram at flat-band conditions for an aluminium –SiO$_2$–p-silicon MOS structure. The doping concentration of the p-silicon is $N_A = 10^{16}$ cm^{-3}, the electron affinity is $W_{EA} = 4.15$ eV, and the metal work function is $W_M = 4.05$ eV.

(b) What gate voltage, V_G, is needed to achieve the flat-band condition?

(c) Repeat (a) and (b) for an n$^+$ polysilicon gate for which the doping

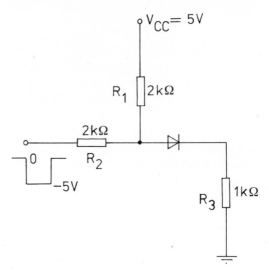

Fig. 1.49. Diode switching circuit.

concentration is just at the edge of degeneracy ($N_D = N_C$).
(Use: $kT/e = 0.025$ V; $n_i = 1.5 \times 10^{10}$ cm^{-3}; effective density of states in the valence band, $N_V = 10^{19}$ cm^{-3}; energy gap $W_g = 1.1$ eV.)

15. The electric field in the SiO$_2$ layer of a MOS capacitor is $E_{ox} = 10^4$ V cm^{-1} and the thickness of the oxide layer is $d_{ox} = 100$ nm.
    (a) What is the space charge in the p-type semiconductor depletion region when the interface charge, Q_Z, can be neglected and electrons at the surface of the semiconductor are absent?
    (b) Calculate the space-charge layer width when the acceptor concentration in the p-type semiconductor is $N_A = 10^{15}$ cm^{-3}.
    (c) Determine the MOS capacitance per unit area for this case.
    (d) What is the space-charge layer width when an electron charge $Q_n'' = 10^{-9}$ As/cm^2 is placed at the p-silicon surface?
    (Use: permittivities $\epsilon_{ox} = 3 \times 10^{-13}$ As/Vcm, $\epsilon_H = 10^{-12}$ As/Vcm; area $A = 100\ \mu m^2$.)

16. Compare the maximum capacitance per unit area of a MOS and p-n junction capacitor (reverse biased) for an operating voltage of 5 V and at maximum fields of $E_{ox} = 10^6$ V/cm and $E_{pn} = 10^5$ V/cm for the MOS capacitor and the p-n junction capacitor, respectively. (Use: $\epsilon_{ox} = 3 \times 10^{-13}$ As/Vcm; $\epsilon_H = 10^{-12}$ As/Vcm; built-in potential, $V_{bi} = 1$ V.)

17. (a) Draw the band diagram of a metal–p-silicon contact. The acceptor concentration is $N_A = 10^{16}$ cm^{-3} (fully ionized) and the hole density at the metal–semiconductor interface is $p(0) = 5 N_A$.
    (b) What is the barrier height, V_B, and the metal work function?

(Use: electron affinity, $W_{EA} = 4.15\,\text{eV}$; energy gap, $W_g = 1.1\,\text{eV}$; intrinsic density, $n_i = 1.5 \times 10^{10}\,\text{cm}^{-3}$; $kT/e = 0.025\,\text{V}$.)

18. Assuming a metal–silicon contact with a barrier height of $V_B = 0.8\,\text{V}$ and a doping concentration of $N_D = 10^{17}\,\text{cm}^{-3}$ donors in the silicon, what is the built-in potential in the semiconductor? (Use: effective density of states in the conduction band, $N_C = 10^{19}\,\text{cm}^{-3}$; $kT/e = 0.025\,\text{V}$.)

19. For a Schottky diode with a metal–semiconductor barrier height $V_B = 0.7\,\text{V}$ and a donor concentration in the n-silicon $N_D = 10^{16}\,\text{cm}^{-3}$, calculate the space-charge layer capacitance per unit area under zero bias conditions. (Use: effective density of states in the conduction band, $N_C = 10^9\,\text{cm}^{-3}$; permittivity $\epsilon_H = 10^{-12}\,\text{As/Vcm}$; electron charge, $e = 1.6 \times 10^{-19}\,\text{As}$; $kT/e = 0.025\,\text{V}$.)

20. Two devices (e.g. MOS transistors) in an integrated circuit are interconnected via a WSi_2 interconnection line with a sheet resistance $R_S = 5\,\Omega$, a length $L = 1\,\text{mm}$, and a width $b = 5\,\mu\text{m}$. The line runs above a $0.5\,\mu\text{m}$ thick SiO_2 layer and is driven by a voltage pulse of $\Delta V = 5\,\text{V}$. At the end of the line a load capacitance $C = 0.2\,\text{pF}$ exists. Find the rise time, t_r, (signal delay) of the voltage v_{out} at the end of the line ($v_{\text{out}}(0) = 0$; $v_{\text{out}}(t_r) = \Delta V/2$). For the interconnection line use a simple T-equivalent circuit model as shown in Fig. 1.50.

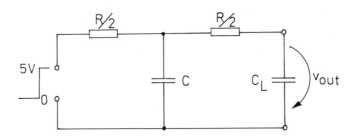

Fig. 1.50. T-equivalent circuit.

2 Transistors

The most important active devices presently available are transistors. They can be used as switches in digital circuits or amplifiers in analogue circuits. There are two main basic types of transistors:

(1) bipolar transistors;
(2) field-effect transistors.

The operational principles of these two types differ considerably. In bipolar transistors both carrier types (electrons and holes) are involved in the operation, whereas in field-effect transistors only one type is important (with the exception of second-order effects). Therefore the field-effect transistors are examples of unipolar transistors. Another difference is that bipolar transistors are current-controlled devices whereas field-effect transistors are voltage controlled.

2.1 BIPOLAR TRANSISTORS

Bipolar transistors are three-terminal active devices consisting of p-n-p or n-p-n structures (Fig. 2.1). The operation is based on minority carrier injection (see Section 1.4.4). The three terminals are emitter (E), base (B), and collector (C).

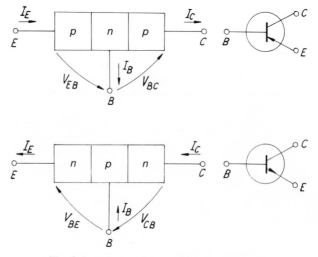

Fig. 2.1. p-n-p and n-p-n bipolar transistors.

2.1.1 Bipolar transistor structures for VLSI

Since the invention of the bipolar transistor in the late 1940s (Shockley 1950) the technology has evolved rapidly. For modern VLSI circuits we need bipolar transistors with very small dimensions in the submicrometre range and super self-aligned structures have been developed (Stevens *et al.* 1985). In Fig. 2.2 we show a cross-section of a typical bipolar transistor for VLSI. Some features and structural details which satisfy the demands of VLSI can be seen in Fig. 2.2 and will be briefly described.

1. To minimize parasitic effects, such as leakage currents and stray capacitors, a good side-wall isolation is necessary. This is accomplished by using deep trenches or grooves of silicon oxide at the borders of the individual devices.

2. For very small lateral dimensions (e.g. emitter width $L_E < 1 \mu$m) and reasonable demands on mask alignment accuracy, self-aligned techniques are employed (Stevens *et al.* 1985). For the example in Fig. 2.2 the emitter doping is carried out by a heavily doped polysilicon layer which is simultaneously the emitter contact. Another self-alignment is given by the SiO_2 grooves as borders of the extrinsic base.

3. In the vertical direction we have very shallow layers for emitter and base ($< 0.5 \mu$m). To minimize spreading resistances (e.g. base resistance $r_{bb'}$) we use high doping concentrations wherever possible (e.g. for emitter and extrinsic base). On the other hand, high doping concentrations cause some unwanted effects such as band narrowing and low breakdown voltages (Del Alama and Swanson 1984; Kumar and Bhat 1989). For small collector spreading resistances, $r_{cc'}$, heavily doped buried n-layers and n-grooves to the collector contact are applied.

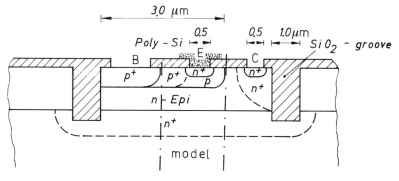

Fig. 2.2. Bipolar transistor for VLSI.

2.1.2 Currents in planar and heteroemitter transistors

For the analysis of transistor operation we employ the planar transistor model shown in Fig. 2.3. which is a very close approximation to a real transistor structure. The n-p-n structure is made by diffusion or ion implantation of donors (n) and acceptors (p) in an n-epitaxy layer. The impurity profile $N_D - N_A = f(x)$, where N_D is the donor concentration and N_A the acceptor concentration, is sketched in Fig. 2.3(b) (see also Section 1.4.1, Equation (1.42)). The emitter and base depth, x_E and x_B, respectively, are some tenths of a micrometre.

First we describe the transistor operation qualitatively. Normally the emitter–base junction is forward biased and the base–collector junction is reverse biased ($V_{BE} > 0$, $V_{BC} < 0$). In this case electrons are injected from the emitter into the base which gives an electron injection current $-I_n(x_E)$. Because of the narrow base these electrons can reach the base–collector junction and are collected there. The loss of electrons in the base due to recombination is very small and we have a collector current

$$I_C = A_N I_E \tag{2.1}$$

where A_N is the common-base current gain. It is somewhat smaller than unity because of the loss of electrons in the base ($I_n(x_E) < I_n(x_B)$) and the contribution of holes to the emitter current ($I_p(x_E)$). We will consider this in more detail later on.

If we control the bipolar transistor with a base current I_B rather than with the emitter current I_E we derive, from Equation (2.1) with $I_E = I_C + I_B$,

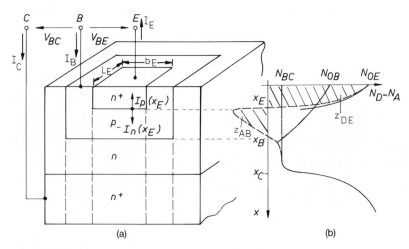

Fig. 2.3. Structure model and impurity profile of an n-p-n bipolar transistor.

$$I_C = \frac{A_N}{1 - A_N} I_B = B_N I_B \tag{2.2}$$

where $B_N = A_N/(1 - A_N)$ is the common-emitter current gain. Let us now go into more detail.

For high enough forward bias (> 0.5 V) the current of the emitter-base diode consists of an electron and a hole injection current only (see Section 1.4.4)

$$I_{ED} = I_n(x_E) + I_p(x_E) . \tag{2.3}$$

If we use Equations (1.76) and (1.78) for the minority injection currents we find

$$I_{ED} = eA_E \left(\frac{D_{nB} n_{iB}^2}{Z_{AB}} + \frac{D_{pE} n_{iE}^2}{Z_{DE}} \right) \left(\exp \frac{V_{BE}}{V_T} - 1 \right) \tag{2.4}$$

or, more generally,

$$I_{ED} = I_{ES} \left(\exp \frac{V_{BE}}{V_T} - 1 \right) . \tag{2.5}$$

$A_E = b_E L_E$ is the emitter area.

For homoemitter transistors the base and emitter intrinsic densities, n_{iB} and n_{iE}, are the same: $n_{iB} = n_{iE} = n_i$. The doping contents in the base and emitter, Z_{AB} and Z_{DE}, are shown in Fig. 2.3(b) as hatched areas. Only the electron injection current into the base, $I_n(x_E)$, is useful for a collector current. Therefore we define an emitter injection efficiency by

$$\gamma_N = \frac{I_n(x_E)}{I_{ED}} = \left(1 + \frac{I_p(x_E)}{I_n(x_E)} \right)^{-1} \tag{2.6}$$

and, with $I_n(x_E)$ and $I_p(x_E)$ from Equation (2.4),

$$\gamma_N = \left(1 + \frac{n_{iE}^2 D_{pE} Z_{AB}}{n_{iB}^2 D_{nB} Z_{DE}} \right)^{-1} . \tag{2.7}$$

γ_N should be unity. Because D_{pE}/D_{nB} is a constant γ_N can be maximized for $Z_{DE}/Z_{AB} \gg 1$ or $n_{iE}/n_{iB} \ll 1$. In classical homoemitter transistors we have $n_{iE} = n_{iB} = n_i$ and the only way to make $\gamma_N \approx 1$ is to ensure $Z_{DE}/Z_{AB} \gg 1$. However, this causes many problems for VLSI transistors.

For VLSI and microwave transistors we have very shallow emitters ($x_E \approx 0.1$–$0.2 \, \mu m$). But the maximum active doping concentration is limited to the order of 10^{20} cm^{-3} because of the maximum solubility of the impurities in the host crystal and the impurity freeze out. Therefore Z_{DE} is limited. For a good emitter injection efficiency γ_N (high current gain) it would be necessary to use small acceptor contents, Z_{AB}, in the base.

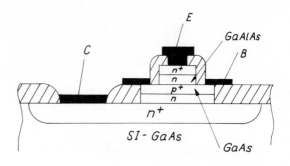

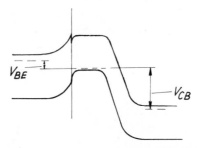

Fig. 2.4. (a) Cross-section and (b) energy-band diagram of a hetero-junction transistor (HBT).

Unfortunately this yields high base spreading resistances $r_{bb'}$. Therefore heteroemitter transistors are often used for high-speed, high-gain microwave transistors (Asbeck *et al.* 1989; Bailbe *et al.* 1987; Hayama *et al.* 1988; Long 1989; Madihan *et al.* 1987; Tiwari and Frank 1989).

In Fig. 2.4 a cross-section of the n-GaAlAs–p-GaAs heterojunction bipolar transistor (HBT) is sketched. Fig. 1.4 shows that GaAlAs has an energy gap, W_g, which is larger than that for GaAs. Therefore the intrinsic density, n_{iE}, in the GaAlAs emitter is smaller than n_{iB} in the GaAs base. If we use x = 0.5 as an example we obtain, from Fig. 1.4, $W_g \approx 1.6\,\text{eV}$. For GaAs we assume $W_g \approx 1.4\,\text{eV}$. Then Equation (1.7) yields

$$\frac{n_{iE}^2}{n_{iB}^2} = \exp\frac{W_{gB} - W_{gE}}{kT} = \exp -8 \approx 3 \times 10^{-4}. \qquad (2.8)$$

In this case we can make $\gamma_N \approx 1$ even for $Z_{DE} < Z_{AB}$. However the process is expensive and complex and there are problems with the materials used for device fabrication.

We will now continue our analysis of the transistor currents. If the loss of electrons due to recombination in the base is I_{rB} the collector transfer current becomes

$$I_{CT} = I_n(x_E) - I_{rB}. \qquad (2.9)$$

The collector transfer current includes all collected electrons coming from the emitter. The total collector current is the sum of this transfer current and a reverse saturation current I_{CB0} due to charge generation in the base–collector space-charge layer

$$I_{CB0} = eA_C \frac{n_i d_{sC}}{\tau_s} \tag{2.10}$$

where A_C is the collector area, d_{sC} is the width of the base–collector space-charge layer, and τ_s is the carrier lifetime in the space-charge layer. We define a base transport factor by

$$\beta_N = \frac{I_{CT}}{I_n(x_E)} = \left(1 + \frac{I_{rB}}{I_{CT}}\right)^{-1} \tag{2.11}$$

Integration of the continuity equation (1.18) for electrons in the base gives the recombination current, I_{rB}, in the base

$$I_{rB} = \frac{Q_B}{\tau_{nB}} \tag{2.12}$$

where Q_B is the charge and τ_{nB} the lifetime of electrons in the base.

The electron transit current through the base is given by (Möschwitzer and Lunze 1990)

$$I_{CT} = \frac{Q_B}{\tau_{BN}} \tag{2.13}$$

where τ_{BN} is the base transit time of electrons. With Equations (2.12) and (2.13) we derive the base transport factor

$$\beta_N = \left(1 + \frac{\tau_{BN}}{\tau_{nB}}\right)^{-1}. \tag{2.14}$$

For a high current gain, β_N should be unity. This means the electrons must travel at high velocity through the base and the transit time in a homogeneously doped base is (Sze 1969)

$$\tau_{BN} = \frac{d_B^2}{2D_{nB}}. \tag{2.15}$$

For example, with $d_B = 0.2 \, \mu m$, $D_{nB} = 20 \, cm^2 \, s^{-1}$ we calculate $\tau_{BN} \approx 10^{-11} \, s$.

With the emitter efficiency γ_N and the base transport factor β_N we obtain the common-base current gain

$$A_N = \frac{I_C}{I_E} = \beta_N \gamma_N \tag{2.16}$$

and finally the common-emitter current gain

$$B_{\mathrm{N}} = \frac{A_{\mathrm{N}}}{1 - A_{\mathrm{N}}} \approx \left(\frac{D_{\mathrm{pE}} Z_{\mathrm{AB}} n_{\mathrm{iE}}^2}{D_{\mathrm{nB}} Z_{\mathrm{DE}} n_{\mathrm{iB}}^2} + \frac{\tau_{\mathrm{BN}}}{\tau_{\mathrm{nB}}} \right)^{-1}. \tag{2.17}$$

The collector transfer current can be expressed as

$$I_{\mathrm{CT}} = A_{\mathrm{N}} I_{\mathrm{ES}} \left(\exp \frac{V_{\mathrm{BE}}}{V_{\mathrm{T}}} - 1 \right) \tag{2.18}$$

In the analysis so far we have considered the transistor to be in 'normal' mode with a forward biased emitter–base junction and a reverse biased base–collector junction. If we change this to the 'inverse' mode of operation with a reverse biased emitter–base junction and a forward biased base–collector junction we obtain qualitatively the same results as for 'normal' mode. For 'inverse' mode the forward current of the base–collector diode can be written as

$$I_{\mathrm{CD}} = I_{\mathrm{CS}} \left(\exp \frac{V_{\mathrm{BC}}}{V_{\mathrm{T}}} - 1 \right) \tag{2.19}$$

and the emitter transfer current is

$$I_{\mathrm{ET}} = A_{\mathrm{I}} I_{\mathrm{CD}} \tag{2.20}$$

where A_{I} is the 'inverse' current again.

The complete current–voltage relations can be obtained by superposition of all currents in normal and inverse modes of operation (Equations (2.5), (2.18)–(2.20))

$$I_{\mathrm{E}} = I_{\mathrm{ED}} - I_{\mathrm{ET}} = I_{\mathrm{ES}} \left(\exp \frac{V_{\mathrm{BE}}}{V_{\mathrm{T}}} - 1 \right) - A_{\mathrm{I}} I_{\mathrm{CS}} \left(\exp \frac{V_{\mathrm{BC}}}{V_{\mathrm{T}}} - 1 \right) \tag{2.21a}$$

$$I_{\mathrm{C}} = I_{\mathrm{CT}} - I_{\mathrm{CD}} = A_{\mathrm{N}} I_{\mathrm{ES}} \left(\exp \frac{V_{\mathrm{BE}}}{V_{\mathrm{T}}} - 1 \right) - I_{\mathrm{CS}} \left(\exp \frac{V_{\mathrm{BC}}}{V_{\mathrm{T}}} - 1 \right) \tag{2.21b}$$

where

$$I_{\mathrm{ES}} A_{\mathrm{N}} = A_{\mathrm{I}} I_{\mathrm{CS}} = I_{\mathrm{S}}. \tag{2.22}$$

2.1.3 Lateral transistors

In an integrated circuit process (e.g. standard buried collector) n-p-n and p-n-p transistors can be made on the same chip using the same process steps. p-n-p lateral transistors can be designed as shown in Fig. 2.5. The current from emitter (E) to collector (C) flows underneath the surface in the lateral (horizontal) direction. The hole injection current of the p-emitter consists of

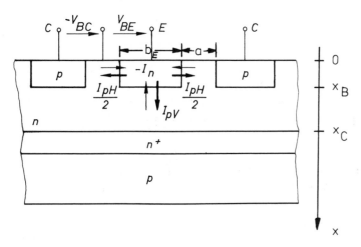

Fig. 2.5. p-n-p lateral transistor.

a horizontal component I_{pH} and a vertical component I_{pV}. The vertical component is useless and has to be minimized for a high current gain.

If we neglect the electron injection current from the base into the emitter the lateral emitter efficiency is given by

$$\gamma_{NL} = \left(1 + \frac{I_{pV} + I_n}{I_{pH}}\right)^{-1} \approx \left(1 + \frac{I_{pV}}{I_{pH}}\right)^{-1} \qquad (2.23)$$

and the base transport factor is (using the same considerations as for planar transistors)

$$\beta_{NL} = \left(1 + \frac{\tau_{BN}}{\tau_p}\right)^{-1} \qquad (2.24)$$

where the lateral hole transit time is

$$\tau_{BN} = \frac{a^2}{2D_p}. \qquad (2.25)$$

τ_p is the life time and D_p the diffusion coefficient of holes. Because of the constant impurity concentration of donors, N_{BC}, in the n-epitaxy layer which is now the base, we have $Z_{DB} = aN_{BC}$ and we derive for the horizontal hole current

$$I_{pH} = eA_{EH}\frac{n_i^2 D_p}{N_{BC}a}\left(\exp\frac{V_{EB}}{V_T} - 1\right). \qquad (2.26)$$

The vertical component is determined by the transport of holes via the n-n$^+$ junction. It is given here without proof (Möschwitzer and Lunze 1990)

$$I_{pV} = eA_{EV} \frac{n_i^2 v_R}{N_{BC}} \left(\exp \frac{V_{EB}}{V_T} - 1 \right) \tag{2.27}$$

where v_R is the recombination velocity at the n-n$^+$ junction and A_{EV} is the vertical emitter area $A_{EV} = b_E L_E$.

The common-base current gain of the lateral transistor is

$$A_{NL} = \gamma_{NL} \beta_{NL} = \left\{ \left(1 + \frac{a^2}{2D_p \tau_p} \right) \left(1 + \frac{b_E v_R a}{2x_B D_p} \right) \right\}^{-1} \tag{2.28}$$

The general current–voltage relations in Equation (2.21) are also valid for lateral transistors.

2.1.4 Polysilicon emitters

Polysilicon emitters (Fig. 2.6) are basically heteroemitters where the hole back-injection current, $I_p(x_E)$, can be decreased and therefore the emitter injection efficiency increased (Ashburn and Soerwidjo 1984; Jalali and Yang 1989; Yu *et al.* 1984). This effect will be analysed in detail using the model in Fig. 2.7.

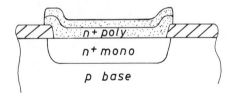

Fig. 2.6. Polysilicon emitter.

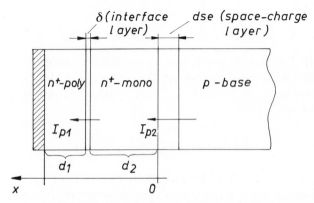

Fig. 2.7. Model of a polysilicon emitter.

We denote the hole back-injection current into the emitter by I_{p2}. In the monocrystalline part of the emitter this is a diffusion current

$$I_{p2} = eA_E D_{p2} \frac{p(0) - p(d_2)}{d_2} \qquad (2.29)$$

where $p(0)$ and $p(d_2)$ are hole boundary concentrations as shown in Fig. 2.7 and D_{p2} is the hole diffusion coefficient in the monocrystalline emitter. Between the monocrystalline and polycrystalline layer we have a very thin oxide layer of thickness ∂. The hole current crossing this layer is a tunnel current

$$I_{pT} = eA_E k_T p(d_2) \qquad (2.30)$$

where

$$k_T = v_{th} P_t / 4 , \qquad (2.31)$$

P_t is the tunnel probability, and v_{th} is the thermal velocity (see Equation (1.113)).

The hole transport current in the polysilicon emitter is again a diffusion current. If the width, d_1, of the polysilicon emitter is much larger than the hole diffusion length, L_{p1}, we find for the hole diffusion current

$$I'_{pt1} = eA_E D_{p1} \frac{p(d_2)}{L_{p1}} . \qquad (2.32)$$

The resulting hole current into the polysilicon emitter is therefore given by

$$\frac{1}{I'_{p1}} = \frac{1}{I_{pT}} + \frac{1}{I'_{pt1}} . \qquad (2.33)$$

The continuity of the hole current flow in the monocrystalline and polycrystalline part of the emitter, $I_{p2} = I_{p1}$, yields a hole boundary concentration.

$$p(d_2) = p(0) \left\{ 1 + \frac{d_2 k_T}{D_{p2}} \left(1 + \frac{L_{p1} k_T}{D_{p1}} \right)^{-1} \right\}^{-1} \qquad (2.34)$$

and, with Equation (2.29),

$$I_{p2} = eA_E D_{p2} \frac{p(0)}{d_2} \left(1 + \frac{D_{p2}}{d_2 k_T} + \frac{L_{p1} D_{p2}}{D_{p1} d_2} \right)^{-1} . \qquad (2.35)$$

On the other hand, if we have no polysilicon layer the hole current would be

$$I_{p2o} = eA_E D_{p2} \frac{p(0) - p(d_2)}{d_2} \approx eA_E D_{p2} \frac{p(0)}{d_2} . \qquad (2.36)$$

The hole back-injection ratio is therefore

$$\frac{I_{p2}}{I_{p2o}} = \left(1 + \frac{D_{p2}}{d_2 k_T} + \frac{L_{p1} D_{p2}}{D_{p1} d_2}\right)^{-1} < 1. \tag{2.37}$$

For polysilicon emitters without any monocrystalline emitter n^+ layer but with tunnel oxide, electron injection into the base occurs by tunnelling (Laser *et al.* 1990). Hole injection into the n-polysilicon emitter can be neglected. Therefore we have

$$\gamma_N \approx 1 \tag{2.38}$$

2.1.5 *I/V characteristics*

The fundamental current–voltage relations for bipolar transistors in Equation (2.21) can be modelled by the Ebers–Moll model in Fig. 2.8. We will consider four operation modes for the n-p-n transistor:

$V_{BE} > 0$ $V_{BC} < 0$ active normal mode;
$V_{BE} < 0$ $V_{BC} > 0$ active inverse mode;
$V_{BE} > 0$ $V_{BC} > 0$ saturation mode;
$V_{BE} < 0$ $V_{BC} < 0$ reverse mode.

For p-n-p transistors the analysis is the same but with opposite signs of currents and voltages.

Ideal characteristics

In the active mode of operation we obtain, from Equation (2.21) with $-V_{BC} \gg V_T$,

$$I_C = A_N I_E + I_{CBO} \tag{2.39}$$

with the collector reverse saturation current

$$I_{CBO} = (1 - A_I A_N) I_{CS}. \tag{2.40}$$

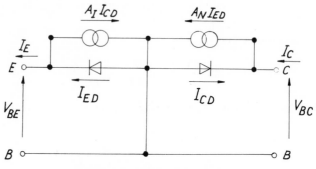

Fig. 2.8. Ebers-Moll model.

The most important circuit configuration for amplifier and switching applications is the common-emitter circuit (Figs 2.9–2.11) because of the small input current.

1. The input characteristics is defined by

$$I_B = f(V_{BE}) \Big|_{V_{CE} = \text{const}}.$$ (2.41)

If we neglect all very small reverse saturation currents in Equation (2.21) we find, with $I_B = I_E - I_C$,

$$I_B = (1 - A_N) I_{ES} \left(\exp \frac{V_{BE}}{V_T} - 1 \right).$$ (2.42)

This characteristic is plotted in Fig. 2.9. It is a diode characteristic with a floating potential $V_{FO}(V_{FO} \approx 0.7\,\text{V}$ for silicon transistors).

2. The current transfer characteristic is defined by

$$I_C = f(I_B) \Big|_{V_{CE} = \text{const}}.$$ (2.43)

For the active normal mode of operation we can employ Equation (2.39) and we obtain, with $I_B = I_E - I_C$,

$$I_C = \frac{A_N}{1 - A_N} I_B + \frac{I_{CBO}}{1 - A_N}$$ (2.44)

or

$$I_C = B_N I_B + I_{CEO}$$ (2.45)

with the common-emitter current gain

$$B_N = \frac{A_N}{1 - A_N}$$ (2.46)

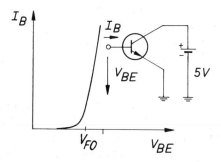

Fig. 2.9. Static input characteristic of a bipolar transistor.

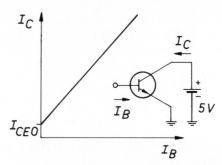

Fig. 2.10. Static current transfer characteristic of a bipolar transistor.

and the common-emitter reverse saturation current

$$I_{CEO} = \frac{I_{CBO}}{1 - A_N} \approx B_N I_{CBO} . \tag{2.47}$$

We sketch this characteristic in Fig. 2.10; it is a straight line with slope B_N.

3. The output characteristic is defined by

$$I_C = f(V_{CE}) \Big|_{I_B = \text{const}} . \tag{2.48}$$

If we substitute $I_B = I_E - I_C$ and $V_{CE} = V_{BE} - V_{BC}$ in Equation (2.21), we have

$$I_C = \frac{B_N I_B \left\{ 1 - \frac{1}{A_I} \exp\left(- \frac{V_{CE}}{V_T}\right) \right\} + I_{CEO} \left\{ 1 - \exp\left(- \frac{V_{CE}}{V_T}\right) \right\}}{1 + \frac{B_N}{B_I} \exp\left(- \frac{V_{CE}}{V_T}\right)} . \tag{2.49}$$

For the active mode of operation, $V_{CE} \gg V_T$, and we derive approximately

$$I_C = B_N I_B + I_{CEO} . \tag{2.50}$$

This means that the I/V plot of Fig. 2.11 is parallel to the V_{CE} axis (thin lines). At low collector voltages, $V_{CE} < V_{FO}$, V_{BC} will be positive and the transistor enters the saturation region. In this case the collector current decreases as can be calculated from Equation (2.49) (see also Fig. 2.11).

The Early effect

For real transistors the collector current in the active mode is not constant. This is also shown in Fig. 2.11 as thick solid lines and is due to the Early effect (Yang 1978). With increasing V_{CE} the base–collector space-charge layer width, d_{sC}, is increased and the electronic base width, $d_B' = d_B - d_{sC}$, is decreased. This yields an increase of the collector current which can be

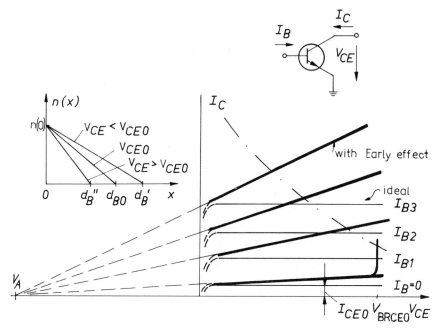

Fig. 2.11. Static output characteristic of a bipolar transistor.

explained by the following simplified considerations. If the emitter–base voltage, V_{BE}, is held constant the electron boundary concentration at the emitter side of the base is also constant (see the inset in the upper left corner of Fig. 2.11):

$$n(0) = n_{\mathrm{op}} \exp \frac{V_{\mathrm{BE}}}{V_{\mathrm{T}}} = \frac{n_i^2 \exp \dfrac{V_{\mathrm{BE}}}{V_{\mathrm{T}}}}{N_A}. \tag{2.51}$$

If we assume that the electron current is a lossless diffusion current throughout the base we have

$$I_C \approx I_E = eA_E D_{nB} \frac{n(0)}{d_B} = eA_E D_{nB} \frac{n_i^2 \exp \dfrac{V_{\mathrm{BE}}}{V_{\mathrm{T}}}}{N_A d_B (V_{\mathrm{CE}})}. \tag{2.52}$$

Because $d_B (V_{CE})$ is a decreasing function of V_{CE}, I_C increases with V_{CE}, as is observed in real transistors (thick solid line in Fig. 2.11).

This effect can be modelled by a linear increase of the collector current with V_{CE}:

$$I_C = (B_N I_B + I_{CEO}) \left(1 + \frac{V_{CE}}{V_A}\right) \tag{2.53}$$

where V_A is an empirical model parameter.

The maximum currents and voltages are limited by the maximum power dissipation

$$I_C V_{CE} = P_{V\text{max}} = \text{const} \qquad (2.54)$$

and the collector breakdown voltage V_{BRCEO} which is mainly due to avalanche effects as we have pointed out in Section 1.4.6.

Temperature effects

The temperature dependence of the collector current can be written as

$$I_C(T) = A_N(T) I_E(T) + I_{\text{CBO}}(T) . \qquad (2.55)$$

It is determined by the temperature dependence of the emitter current, $I_E(T)$, the collector reverse saturation current, $I_{\text{CBO}}(T)$, and the current gain, $A_N(T)$. In Fig. 2.12 a measured temperature dependence of the common-emitter current gain $B_N(T)$ is shown. It can be approximated by

$$B_N(T) = B_N(T_o) \exp 0.006(T - T_o) . \qquad (2.56)$$

From this we obtain the temperature dependence of the common-base current gain

$$A_N(T) = \frac{B_N(T)}{1 + B_N(T)} . \qquad (2.57)$$

Much more temperature dependent are the currents I_E and I_{CBO}. I_{CBO} is basically a reverse current. In silicon devices it is proportional to the intrinsic density, n_i. Therefore we have an exponential temperature dependence

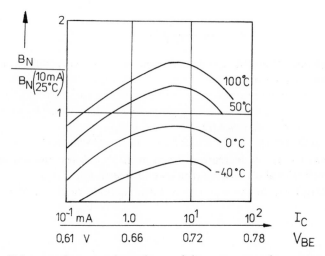

Fig. 2.12. Voltage and current dependence of the common-emitter current gain.

$$I_{CBO}(T) = I_{CBO}(T_o) \exp c_r(T - T_o) \tag{2.58}$$

where

$$c_r = \frac{W_t}{kT_o^2} = \frac{W_g}{mkT_o^2} \tag{2.59}$$

and W_t is the energy level of the recombination centres (Möschwitzer and Lunze 1990). The emitter current, I_E, is a forward current. Because of the temperature dependence $\exp eV_{BE}/kT$ and Equation (1.107) in Section 1.4.8, we derive

$$I_E = I_E(T_o) \exp c_{fe}(T - T_o) \tag{2.60}$$

where

$$c_{fe} = \frac{W_g - eV_{BE}}{kT_o^2} \tag{2.61}$$

where W_g is the energy gap (see Chapter 1). The inner temperature T of the transistor is determined by the electric power $P_V = V_{CE}I_C$ and the total thermal resistance R_{th}:

$$T = T_u + R_{th}P_V \tag{2.62}$$

where T_u is the ambient temperature. The thermal resistance is determined by the thermal conductivity and the heat dissipation by convection at the surface (for more details see Section 5.2).

To describe the temperature dependence it is useful to define a temperature feed-through by

$$D_T = \frac{\Delta V_{BE}}{\Delta T}\bigg|_{I_C=\text{const}}. \tag{2.63}$$

This is the change of the emitter–base voltage ΔV_{BE} which is needed to keep the collector current I_C constant for a temperature variation ΔT. It is negative. The temperature dependence of the collector current and the large shift of the operating point is shown in Fig. 2.13 with $V_{BE} = 0.6\,\text{V}$, I_C (293 K) = 5 mA, and $c_{fe} = 0.078$.

To keep the operating point nearly constant, special circuit techniques are employed; a simple solution with an emitter resistor, R_E, is shown in Fig. 2.20. We find for the temperature feed-through

$$D_T = \frac{\Delta V_{BE}}{\Delta T}\bigg|_{I_C} = -\frac{1}{T_o}\left(\frac{W_g}{e} - V_{BE}\right). \tag{2.64}$$

A numerical example with $V_{BE} = 0.6\,\text{V}$; $T_o = 300\,\text{K}$; $W_g = 1.1\,\text{eV}$ gives $D_T = -1.7\,\text{mV K}^{-1}$.

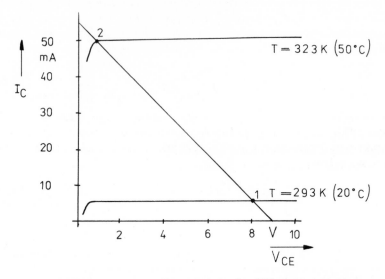

Fig. 2.13. Collector current characteristics for different temperatures.

Because of thermal feedback a thermal runaway can occur. This runaway can be prevented if the rate of dissipated power, dP_V/dT, is larger than the rate of electrical power, dP_e/dT. Using Equation (2.62) for dP_V/dT yields

$$\frac{dP_e}{dT} < \frac{1}{R_{th}} \qquad (2.65)$$

where

$$\frac{dP_e}{dT} = V_{CE}\frac{dI_C}{dT} + I_C\frac{dV_{CE}}{dT}. \qquad (2.66)$$

The ideal case, $dP_e/dT = 0$, is derived by

$$V_{CE}dI_C = -I_CdV_{CE} \qquad (2.67)$$

or

$$\frac{V_{CE}}{I_C} = -\frac{dV_{CE}}{dI_C} = R_C \qquad (2.68)$$

where R_C is the load resistor in the collector branch (see Fig. 2.20). In other words, Equation (2.68) means that the voltage drops across the load resistor and the transistor must be equal. Unfortunately this cannot be realized in every case.

Second-order effects

Current dependence of the current gain   Fig. 2.12 shows a decrease of the current gain at low and high collector currents. For low currents (low V_{BE})

the decrease of the current gain is due to the increasing effect of the use-less recombination current in the emitter–base space-charge layer (see Equation (1.66)).

$$I_{r_gE} = eA_E \frac{n_i d_{sE}}{\tau_s} \exp \frac{V_{BE}}{2V_T}.$$ (2.69)

The decrease of the current gain at high currents is due to high-injection effects (Möschwitzer and Lunze 1990; Muller and Kamins 1977; Yang 1978). At high current densities the majority carrier concentration will be increased: $n = n_i \exp V_{BE}/2V_T$ and in the current equation (2.4) we now have to use

$$Z_{AB} + \frac{n_i d_B}{2} \exp \frac{V_{BE}}{2V_T}$$

instead of Z_{AB}.

In summary, we derive an emitter injection efficiency including space-charge layer recombination and high-injection effects:

$$\gamma_N = \left\{ 1 + \frac{\left(Z_{AB} + \frac{n_i}{2} d_B \exp \frac{V_{BE}}{2V_T}\right)\left(D_{pE} + \frac{d_{sE}}{n_i \tau_s} Z_{DE} \exp \frac{-V_{BE}}{2V_T}\right)}{Z_{DE} D_{nB}} \right\}^{-1}$$ (2.70a)

and especially for $\beta_N \approx 1$, $A_N \approx \gamma_N$, and $n_{iE} = n_{iB}$ (a homoemitter):

$$B_N = \frac{\gamma_N}{1 - \gamma_N} = \frac{D_{nB} Z_{DE}}{D_{pE} Z_{AB} \left\{ \frac{n_i d_B}{2Z_{AB}} \exp \frac{V_{BE}}{2V_T} + 1 \right\} \left\{ \frac{Z_{DE} d_{sE}}{n_i \tau_s D_{pE}} \exp\left(-\frac{V_{BE}}{2V_T}\right) + 1 \right\}}.$$ (2.70b)

This equation explains qualitatively the plots in Fig. 2.12.

Influence of spreading resistors   The most important effect of spreading resistors is due to the base resistor, $r_{bb'}$. Because of the lateral voltage drop on this resistor we observe the pinch effect and the current-crowding effect.

The pinch effect is explained in Fig. 2.14. It can be initiated by an avalanche current at the base–collector junction. In this case a lateral base current flows from the internal base to the outside base contact causing a base–emitter forward bias, V_{BE}, that is higher in the centre of the emitter than at the periphery. Because of the exponential dependence, $I \sim \exp V_{BE}/V_T$, the current is pinched to a thin filament near the centre of the emitter. This yields a local temperature rise and instabilities (second breakdown (Yang 1978)).

In normal modes of operation the base current flows from the outside base contacts to the inner base region in lateral directions as shown in Fig. 2.15. In this case the base–emitter forward bias, V_{BE}, at the periphery of the

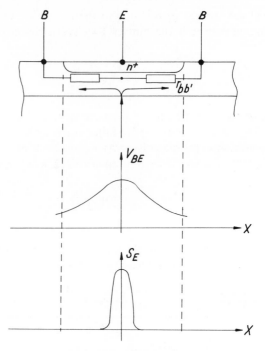

Fig. 2.14. Pinch effect.

emitter is higher than in the centre. This leads to current crowding at the emitter periphery. Therefore a design goal is to make the ratio of emitter periphery to emitter area as large as possible (see also Section 2.1.10).

Taking all these effects into account we obtain an approximate expression for the optimal emitter current per emitter length L_E. It is given here without proof (Möschwitzer and Lunze 1990):

$$I_E/L_E = V_T\sqrt{2B_N}\,\frac{1}{R_{SBi}d_B}$$ (2.71)

where R_{SBi} is the sheet resistance of the inner base (e.g. $R_{SBi} = 1{,}000$–$2{,}000\,\Omega/\square$) and $V_T = kT/e$

2.1.6 Network models

Large-signal model

A widely used network model for computer simulation of bipolar transistor circuits is shown in Fig. 2.16. This can be obtained very easily from the basic Ebers–Moll model in Fig. 2.8 by using the following currents

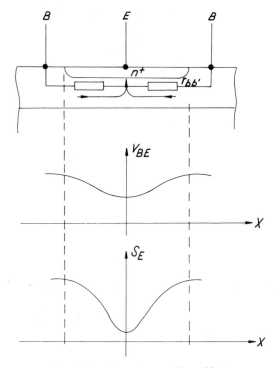

Fig. 2.15. Current-crowding effect.

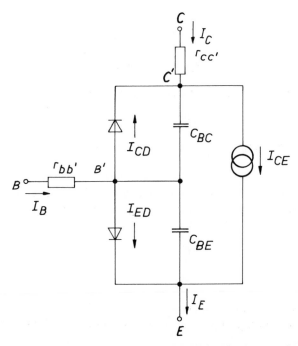

Fig. 2.16. Non-linear network model of the bipolar transistor.

$$I_{CD} = \frac{I_S}{B_1}\left(\exp\frac{V_{B'C'}}{V_T} - 1\right) \tag{2.72}$$

$$I_{ED} = \frac{I_S}{B_N}\left(\exp\frac{V_{B'E}}{V_T} - 1\right) \tag{2.73}$$

$$I_{CE} = I_S\left(\exp\frac{V_{B'E}}{V_T} - \exp\frac{V_{B'C'}}{V_T}\right) \tag{2.74a}$$

$$I_S = A_N I_{ES} = A_1 I_{CS} \tag{2.75}$$

or including the Early effect (see Section 2.1.5; Equation (2.53))

$$I_{CE} = I_S\left(\exp\frac{V_{B'E}}{V_T} - \exp\frac{V_{B'C'}}{V_T}\right)\left(1 + \frac{V_{C'E}}{V_A}\right). \tag{2.74b}$$

The base and collector spreading resistances $r_{bb'}$ and $r_{cc'}$ cause a voltage drop between the external terminals and the inner base and collector nodes, respectively.

$$V_{BE} = V_{B'E} + I_B r_{bb'} \tag{2.76}$$

$$V_{CE} = V_{C'E} + I_C r_{cc'}. \tag{2.77}$$

In summary we have the static model parameters

$$I_S, B_N, r_{bb'}, r_{cc'}, B_1, V_A$$

which can be measured or calculated by using the results of Sections 2.1.2–2.1.4.

For dynamic simulations the model must be supplemented with the base–collector and base–emitter capacitors, C_{BC} and C_{BE}, respectively. These capacitors consist of space-charge layer capacitors, C_{sc} and C_{se} and diffusion capacitors, C_{dc} and C_{de}:

$$C_{BC} = C_{sc} + C_{dc} \tag{2.78}$$

$$C_{BE} = C_{se} + C_{de}. \tag{2.79}$$

Employing the results of Section 1.4.3. (Equation (1.63)) for p-n junctions we obtain the voltage dependencies of the collector and emitter space-large layer capacitances

$$C_{sc} = C_{sco}\left(1 - \frac{V_{B'C'}}{V_{bi}}\right)^{-q}$$ (2.80)

$$C_{se} = C_{seo}\left(1 - \frac{V_{B'E}}{V_{bi}}\right)^{-q}$$ (2.81)

where $V_{bi} = 0.8-1$ V; $q = 0.3-0.5$.

The diffusion capacitances can be estimated from the minority carrier charge stored in the base of a transistor in normal and inverse mode, Q_{BN} and Q_{BI}, respectively. For these minority carrier charges we have

$$Q_{BN} = \tau_{BN} I_S \exp \frac{V_{B'E}}{V_T}$$ (2.82)

$$Q_{BI} = \tau_{BI} I_S \exp \frac{V_{B'C'}}{V_T}$$ (2.83)

where τ_{BN} and τ_{BI} are the normal and inverse base transit times, respectively. We find the diffusion capacitances

$$C_{de} = \frac{dQ_{BN}}{dV_{B'E}} = \frac{\tau_{BN} I_S}{V_T} \exp \frac{V_{B'E}}{V_T} = C_{deo} \exp \frac{V_{B'E}}{V_T}$$ (2.84)

$$C_{dc} = \frac{dQ_{BI}}{dV_{B'C'}} = \frac{\tau_{BI} I_S}{V_T} \exp \frac{V_{B'C'}}{V_T} = C_{dco} \exp \frac{V_{B'C'}}{V_T}.$$ (2.85)

C_{seo}, C_{sco}, C_{deo}, C_{dco} are four additional model parameters which should be measured.

Small-signal model

The small-signal model is very important for the analysis and simulation of analogue transistor circuits. We consider the active normal mode of operation and regard the small-signal a.c. currents and voltages as small differences or differentials of the large-signal d.c. currents and voltages (e.g. $i_b = dI_B$, $v_{be} = dV_{BE}$). We then derive from Fig. 2.16 the small-signal model in Fig. 2.17 where we have neglected $r_{cc'}$.

For the active mode of operation the d.c. base current is

$$I_B = \frac{I_S}{B_N} \exp \frac{V_{B'E}}{V_T}$$ (2.86)

and the small-signal input resistance, r_e in Fig. 2.17, will be

$$r_e = \frac{dV_{B'E}}{dI_B} = \frac{V_T}{I_B} \approx \frac{V_T}{I_E} B_N = r_{de} B_N.$$ (2.87)

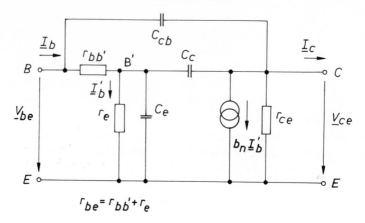

Fig. 2.17. Small-signal model of the bipolar transistor.

The d.c. collector transfer current is

$$I_{CE} = I_s \exp \frac{V_{B'E}}{V_T} = B_N I_B \qquad (2.88)$$

and we obtain

$$i_{ce} = dI_{CE} = B_N dI_B + dB_N I_B . \qquad (2.89)$$

If we assume a constant d.c. current gain B_N and designate the common-emitter small-signal current gain as b_n, so that $dI_C/dI_B = B_N = b_n$, the small-signal current source becomes

$$i_{ce} = b_n i_b . \qquad (2.90)$$

The a.c. small-signal output resistance is

$$r_{ce} = \frac{dV_{C'E}}{dI_C} = \frac{V_A}{I_C} . \qquad (2.91)$$

For example, for $V_A = 100\,\text{V}$ and $I_C = 1\,\text{mA}$ we have $r_{ce} = 100\,\text{k}\Omega$.
For the capacitors C_e and C_c we have

$$C_e = C_{se} + C_{de} \qquad (2.92)$$

$$C_c = C_{sc} . \qquad (2.93)$$

C_c is determined by the space-charge layer capacitance only because in the active normal mode of operation the collector diffusion capacitance is zero.

This small-signal model can be further extended by adding stray capacitors and inductors. Because of its importance only the collector–base stray capacitance, C_{cb}, is added in Fig. 2.17. It should be mentioned that in

Fig. 2.17 for all a.c. currents and voltages the complex phasors $\mathbf{V}$ and $\mathbf{I}$ are used instead of the real time varying functions i and v.

2.1.7 High-frequency behaviour

The high-frequency behaviour is determined by the frequency response of the two-port parameters. For bipolar transistors the following two-port equations are widely used:

(1) hybrid (h)-equations

$$\mathbf{V}_{be} = h_{11e}\mathbf{I}_b + h_{12e}\mathbf{V}_{ce} \tag{2.94}$$

$$\mathbf{I}_c = h_{21e}\mathbf{I}_b + h_{22e}\mathbf{V}_{ce}, \tag{2.95}$$

(2) y-equations

$$\mathbf{I}_b = y_{11e}\mathbf{V}_{be} + y_{12e}\mathbf{V}_{ce} \tag{2.96}$$

$$\mathbf{I}_c = y_{21e}\mathbf{V}_{be} + y_{22e}\mathbf{V}_{ce}. \tag{2.97}$$

The y-parameters y_{ik} can be derived from the h-parameters and vice versa (Möschwitzer and Lunze 1990). Some interesting formulas are

$$y_{11} = 1/h_{11}$$

$$y_{21} = h_{21}/h_{11}$$

$$h_{21} = y_{21}/y_{11}.$$

Let us now consider the frequency response of some h- and y-parameters. For simplicity, we initially neglect C_c, C_{cb}, and r_{ce} without significant loss of accuracy.

1. common-emitter current gain

$$h_{21e} = \left.\frac{\mathbf{I}_c}{\mathbf{I}_b}\right|_{\mathbf{V}_{ce}=0} = -\frac{b_n}{1 + j\omega C_e r_e}. \tag{2.98}$$

The frequency response of h_{21e} is shown in Fig. 2.18. From this plot we can see two characteristic frequencies: the common-emitter cut-off frequency, f_β, is defined by

$$|h_{21e}(f_\beta)| = \frac{b_n}{\sqrt{2}} \tag{2.99}$$

and, with h_{21e} from Equation (2.98), we find

$$f_\beta = \frac{1}{2\pi C_e r_e}. \tag{2.100}$$

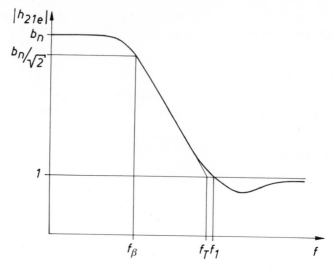

Fig. 2.18. Common-emitter current gain as a function of frequency.

A further characteristic frequency is the transit frequency, f_T, or gain–bandwidth product which is defined by

$$|h_{21e}(f)| f = f_T. \tag{2.101}$$

Using Equation (2.98) we get

$$f_T = \frac{\sqrt{b_n^2 - 1}}{2\pi C_e r_e} \approx b_n f_\beta. \tag{2.102}$$

Because of the influence of C_{cb} at very high frequencies there is a difference between f_T and that frequency f_1 for which the $|h_{21e}|$ is unity. This is also sketched in Fig. 2.18. The transit frequency f_T can be calculated using the small-signal equivalent circuit in Fig. 2.17

$$\frac{1}{2\pi f_T} = \frac{C_e r_e}{b_n} = C_e r_{de} = C_{se} r_{de} + C_{de} r_{de}. \tag{2.103}$$

$\tau_E = C_{se} \, r_{de}$ is the emitter loading time and $\tau_{BN} = C_{de} \, r_{de}$ is the base transit time ($r_{de} = V_T/I_E$, see Equation (2.87)). If we consider the collector space-charge layer capacitance, $C_c = C_{sc}$, and the collector transit time, τ_{CT}, as well, we obtain a more precise expression for the transit frequency

$$\frac{1}{2\pi f_T} = \tau_E + \tau_{BN} + \tau_{CT} + r_{de} C_{sc}. \tag{2.104}$$

f_T increases with increasing emitter current ($f_T \sim 1/r_{de} = I_E/V_T$ (see Equations (2.103) and (2.104)) but at high currents f_T decreases with increasing emitter current because of the Kirk effect (Möschwitzer and Lunze 1990; Muller and Kamins 1977; Yang 1978): At high current densities the effective base width and therefore the base transit time, τ_{BN}, will be increased because of high-injection effects.

2. The transconductance is

$$y_{21e} = \frac{I_c}{V_{be}}\bigg|_{V_{ce}=0} = \frac{b_n}{r_e\left(\dfrac{r_{bb'}}{r_e} + j\omega C_e r_{bb}' + 1\right)} \tag{2.105}$$

or

$$y_{21e} = \frac{1}{r_{de}\left(1 + j\omega C_e r_{bb}' + \dfrac{r_{bb'}}{r_e}\right)} = \frac{I_E}{V_T\left(1 + j\omega C_e r_{bb}' + \dfrac{r_{bb'}}{r_e}\right)}. \tag{2.106}$$

For low frequencies we have, with $r_{bb}' \ll r_e$,

$$y_{21eo} = g_m = \frac{1}{r_{de}} = \frac{I_E}{V_T}. \tag{2.107}$$

3. The input resistance is

$$h_{11e} = \frac{1}{y_{11e}} = \frac{V_{be}}{I_b}\bigg|_{V_{ce}=0} = r_{bb'} + \frac{r_e}{1 + j\omega C_e r_e}. \tag{2.108}$$

At very high frequencies we have $r_e \| C_e \ll r_{bb'}$ and therefore

$$h_{11e} \approx r_{bb'} \quad \text{and} \quad |h_{21e}| \approx \frac{\omega_T}{\omega}.$$

The maximum available small-signal power amplification is

$$v_{pam} = \frac{\omega_T}{4\omega^2 r_{bb'} C_c}. \tag{2.109}$$

This will be unity at a frequency f_{max} given by

$$f_{max} = \sqrt{\frac{f_T}{8\pi r_{bb'} C_c}} \tag{2.110}$$

It should be mentioned that for $f < f_{max}$ the bipolar transistor is an active device. To get a large f_{max} the product $r_{bb'} C_c$ must be minimized and this can be done by special transistor designs as shown in Fig. 2.19. For an elementary microwave transistor, like that in Fig. 2.19, we have

$$r_{bb'} C_c = r_k(C_a + C_s + C_{sc}) + (r_{bb'1} + r_{bb'2})C_{sc}. \tag{2.111}$$

$r_{bb'1}$ is minimized by using narrow emitters and interdigital structures;

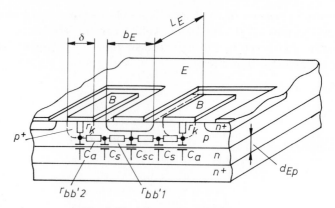

Fig. 2.19. Microwave transistor.

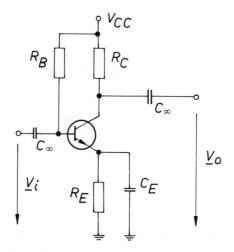

Fig. 2.20. Common-emitter amplifier stage.

$r_{bb'2}$ can be decreased by heavy doping of the extrinsic base. The contact resistance, r_k, is kept low by employing carefully selected contact materials. To get small space-charge layer capacitances C_a, C_s, and C_{sc} the areas and the epitaxy layer doping is kept as low as possible.

2.1.8 Small-signal amplifier stages

A simple common-emitter amplifier stage is shown in Fig. 2.20. The operating point is determined by the resistors R_B, R_E, and R_C:

$$I_B = \frac{V_{CC} - V_{F0}}{R_B + R_E B_N}. \tag{2.112}$$

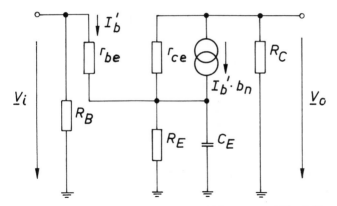

Fig. 2.21. Equivalent circuit of the amplifier stage of Fig. 2.20.

where, for example, $V_{F0} = 0.7\,\text{V}$ for silicon transistors. R_E is used for negative serial feedback to stabilize the operating point. It must be shunted by a large capacitor C_E. The coupling capacitors, C_∞, can be considered as infinite in this case; they have no influence on our considerations.

Using the small-signal equivalent circuit in Fig. 2.17 we find the low-frequency equivalent circuit of the amplifier stage in Fig. 2.21. In the low-frequency range now under consideration the inner capacitors, C_e and C_c, are neglected and $r_{be} = r_{bb'} + r_e$ is used. We may set $r_{ce} \to \infty$ and find a voltage gain

$$v_u = \frac{\mathbf{V}_o}{\mathbf{V}_i} = \frac{-R_C b_n}{r_{be} + (1 + b_n)\dfrac{R_E}{1 + j\omega C_E R_E}} . \qquad (2.113)$$

At extremely low frequencies we have a drift gain

$$v_{uo} = -\frac{R_C b_n}{r_{be} + (1 + b_n) R_E} \approx -\frac{R_C}{R_E} . \qquad (2.114)$$

At reasonable signal frequencies, $\omega \gg 1/C_E R_E$ and we have

$$v_u = -\frac{R_C b_n}{r_{be}} \approx -\frac{R_C}{r_{de}} = -\frac{R_C I_C}{V_T} . \qquad (2.115)$$

The small-signal voltage gain is therefore determined mainly by the d.c. collector current, I_C, and the collector load resistor, R_C.

The differential amplifier stage in Fig. 2.22 is a basic cell for integrated analogue circuits. The differential amplification is

$$v_D = \frac{\mathbf{V}_{o2} - \mathbf{V}_{o1}}{\mathbf{V}_{i2} - \mathbf{V}_{i1}} = -\frac{b_n R_C}{r_{be}} \qquad (2.116)$$

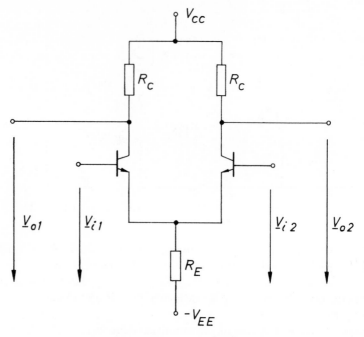

Fig. 2.22. Differential amplifier stage.

and the common-mode amplification is

$$v_C = \frac{\mathbf{V}_o}{\mathbf{V}_i} = -\frac{R_C b_n}{2R_E(1 + b_n) + r_{be}} \qquad (2.117)$$

where $\mathbf{V}_o = \mathbf{V}_{o1} = \mathbf{V}_{o2}$ and $\mathbf{V}_i = \mathbf{V}_{i1} = \mathbf{V}_{i2}$.

2.1.9 Switching behaviour

In digital circuits the transistor is used as a switch and Fig. 2.23 shows the basic circuit for a transistor switch. The transistor is switched on by a positive base current, I_{B1}, if the pulse voltage, V_B, at the input is high enough:

$$I_{B1} = \frac{R_E V_B - R_B V_{EE} - (R_B + R_E) V_{FO}}{R_B R_E}. \qquad (2.118)$$

It is switched off if $V_B = 0$ and a negative base current $-I_{B2}$ is supplied:

$$I_{B2} = -\frac{R_B V_{EE} + (R_B + R_E) V_{FO}}{R_B R_E}. \qquad (2.119)$$

Equation (2.119) is true as long as the base–emitter junction is forward

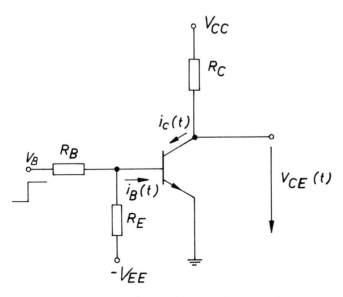

Fig. 2.23. Bipolar transistor switching circuit.

biased (conducting). If the base–emitter junction becomes reverse biased, I_{B2} is nearly zero.

This switching operation can also be explained by means of the output characteristic shown in Fig. 2.24. The intersection of the I/V characteristic of the load resistor, R_C, with the output characteristic of the transistor gives the two operation points in the 'on' and 'off' state. The maximum possible collector current is determined by the power supply voltage, V_{CC}, and the load resistor, R_C:

$$I_{Cm} = \frac{V_{CC}}{R_C}. \tag{2.120}$$

More precisely, taking the small saturation voltage, V_{CES}, in the on state of the transistor into account we have, for the maximum collector current,

$$I_{C1} = \frac{V_{CC} - V_{CES}}{R_C} \approx I_{Cm}. \tag{2.121}$$

To make the collector current independent of the current gain and the base current, the transistor switch must be driven into saturation. This could happen if

$$B_N I_{B1} \gg \frac{V_{CC}}{R_C}. \tag{2.122}$$

The collector–emitter saturation voltage is the sum of the internal

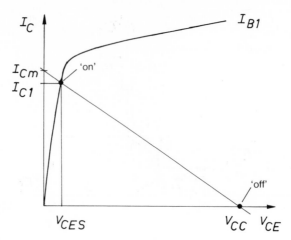

Fig. 2.24. Collector current characteristic showing the on and off states of the switching transistor.

collector–emitter voltages $V_{C'ES}$ and the voltage drop on the collector spreading resistor

$$V_{CES} = I_{C1} r_{cc'} + V_{C'ES}. \qquad (2.123)$$

Using Equation (2.49) we find the collector–emitter saturation voltage to be

$$V_{C'ES} = -V_T \ln \frac{-1 + m}{\dfrac{B_N}{B_I} + \dfrac{m}{A_I}} \qquad (2.124)$$

where

$$m = \frac{B_N I_{B1}}{I_{C1}} \gg 1$$

is the overdriving factor

The time response of the transistor switch is shown in Fig. 2.25 for a step function of the base current $i_B(t)$. The response of the collector current is delayed because of the transit time and the time needed for charging and discharging the transistor. This can be analysed with the charge-control equation (Möschwitzer and Lunze 1990)

$$i_B = \frac{Q_B}{\tau_{nB}} + \frac{dQ_B}{dt} + \frac{dQ_S}{dt}. \qquad (2.125)$$

This equation can be understood as follows. The base current i_B is used to change the minority carrier charge in the base, dQ_B/dt, and in the base–collector space-charge layer, dQ_S/dt. It is also used to deliver minority carriers

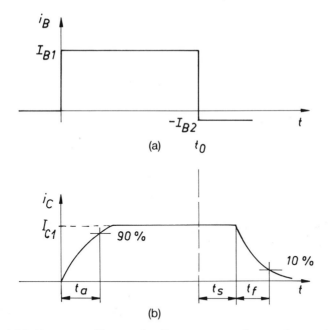

Fig. 2.25. Response of base and collector current of a transistor switch.

for recombination in the base, Q_B/τ_{nB}, where τ_{nB} is the minority carrier lifetime in the base (see also Section 2.1.2, Equation (2.12)). If we employ Equation (2.13) the relation between collector current, i_C, and base charge, Q_B, is

$$i_C = \frac{Q_B}{\tau_{BN}} \tag{2.126}$$

where τ_{BN} is the base transit time. To be precise Equation (2.126) is only valid for steady-state conditions.

Furthermore, we derive from Fig. 2.23

$$\frac{dQ_S}{dt} = C_{sc}\frac{dv_{BC}}{dt} = R_C C_{sc}\frac{di_C}{dt}. \tag{2.127}$$

Equations (2.126) and (2.127) give

$$i_B = (\tau_{BN} + R_C C_{sc})\frac{di_C}{dt} + \frac{\tau_{BN}}{\tau_{nB}}i_C. \tag{2.128}$$

Using the results of Section 2.1.2 Equation (2.17) it can be shown that we have $B_N \approx \tau_{nB}/\tau_{BN}$ for $\gamma_N \approx 1$. Therefore Equation (2.128) can be simplified to

$$B_N i_B = \tau_a \frac{di_C}{dt} + i_C \tag{2.129}$$

where

$$\tau_a = \tau_{nB} + B_N R_C C_{sc} \tag{2.130}$$

is a characteristic time constant.

Now let us consider the time response of the collector current in the active normal mode for which Equation (2.129) is valid. Solving Equation (2.129) with $i_B = I_{B1}$ and the initial condition $i_C(0) = 0$ (see Fig. 2.25(b)) yields

$$i_C(t) = B_N I_{B1} \left[1 - \exp\left(-\frac{t}{\tau_a}\right) \right] \tag{2.131}$$

which is valid only for $i_C \le I_{C1}$. The rise time t_a is defined by $i_C(t_a) = 0.9 I_{C1}$ and is found, with Equation (2.131), to be

$$t_a = \tau_a \ln \frac{m}{m - 0.9} \tag{2.132}$$

where

$$m = \frac{B_N I_{B1}}{I_{C1}} \gg 1. \tag{2.133}$$

If at $t = t_o$ the base current is switched to $-I_{B2}$ (see Fig. 2.25) the collector current stays constant for the storage time, t_s. During this storage time, the excess charge, Q_{BS}, stored in the base in the saturation mode is removed. This excess charge Q_{BS} in the saturation mode is due to an excess base current $i_B - I_{C1}/B_N$ and can be calculated with the charge-control equation (Möschwitzer and Lunze 1990)

$$i_B - \frac{I_{C1}}{B_N} = \frac{dQ_{BS}}{dt} + \frac{Q_{BS}}{\tau_s} \tag{2.134}$$

where τ_s is the storage time constant. It can be considered as a given parameter.

For $t > t_o$ we have $i_B = -I_{B2}$. Solving Equation (2.134) with the initial condition $Q_{BS}(t_o) = \tau_s(I_{B1} - I_{C1}/B_N)$ we obtain

$$Q_{BS}(t) = \tau_s(I_{B1} + I_{B2}) \exp\left\{ -\left(\frac{t - t_o}{\tau_s}\right) \right\} - \tau_s\left(I_{B2} + \frac{I_{C1}}{B_N}\right). \tag{2.135}$$

The storage time is defined by $Q_{BS}(t - t_o = t_s) = 0$, that is, when the excess charge has disappeared. If we set, in Equation (2.135), $Q_{BS} = 0$ and $t - t_o = t_s$ we find

$$t_s = \tau_s \ln \frac{m + k}{1 + k} \tag{2.136}$$

where m is derived from Equation (2.133) and k is

$$k = \frac{B_N I_{B2}}{I_{C1}}.$$ (2.137)

For $t > t_o + t_s$ the transistor enters the active region and the charge-control equation (2.129) is again valid. Solving Equation (2.129) with $i_B = -I_{B2}$ and i_C ($t = t_o + t_s) = I_{C1}$ (see Fig. 2.25(b)) yields the time response of the collector current

$$i_C = (I_{C1} + B_N I_{B2}) \exp \left[-\left(\frac{t - t_o - t_s}{\tau_a} \right) \right] - B_N I_{B2}.$$ (2.138)

The fall time, t_f, is defined by i_C $(t - t_o - t_s = t_f) = 0.1 I_{C1}$ (see Fig. 2.25(b)) and we find, with Equation (2.138)

$$t_f = \tau_a \ln \frac{k + 1}{k + 0.1}$$ (2.139)

where k can be derived from Equation (2.137).

The signal delay, t_s, due to the excess minority charge, Q_{BS}, stored during saturation is a disadvantage. It can be avoided by clamping the collector–base junction by a Schottky diode (see Section 1.5.2). In Fig. 2.26 we sketch a cross-section of a Schottky-clamped bipolar transistor used in

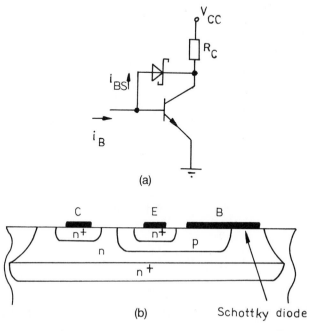

Fig. 2.26. Bipolar transistor clamped with a Schottky diode.

integrated circuits. The Al–n-silicon junction forms a Schottky diode and
the Al–p-silicon junction is an ohmic contact.

2.1.10 Physical implementation (layout)

In Fig. 2.27 we show a layout (a) and a cross-section (b) of a bipolar
transistor fabricated in standard buried collector technology (SBC). SBC
and its derivatives are currently the main means of batch production of standard bipolar ICs and dictate the vertical sheet structure of those devices. The
n-p-n transistors are produced in an n-epitaxy layer which is grown on a p-
type substrate. The side-wall isolation of the active transistor areas is
achieved by doping deep p-type diffused grooves. In this classical SBC structure we have therefore an isolation by space-charge layers of reverse biased
p-n junctions. By diffusion or ion implantation of p- and n-layers the base
and the emitter regions are formed. The buried n^+-layer underneath the n-
epitaxy layer reduces the collector spreading resistance $r_{CC'}$.

The layout (Fig. 2.27(a)) consists of several mask levels for the process
steps: active transistor areas A; base area B; emitter area E; contact windows
C; and interconnection lines D. The design engineer (draughtsman) must
observe design rules for these levels, such as minimum feature sizes and

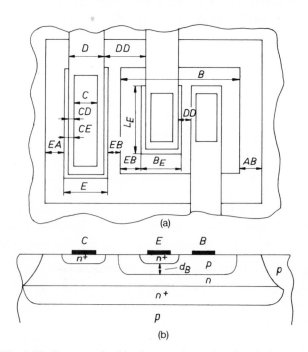

Fig. 2.27. Layout of a bipolar transistor showing design rules.

overlaps of the individual mask levels. The design rules are determined by physical and lithographic constraints.

1. The structures (polygons) are specified by a lithographic process. The classical process is photolithography using ultraviolet light and minimum feature sizes down to 1 μm (or somewhat below) are possible. More recently electron beam lithography is also employed for feature sizes in the submicrometre range. The history of minimum possible feature sizes can be roughly summarized as follows

1965	10 μm
1975	5 μm
1985	1 μm
1995	0.5 μm

2. The physical reasons for restricted design rules are breakdown and parasitic effects which prevent proper operation of the device.

In Fig. 2.27 some design rules are depicted:

(1) B minimum size of the base regions;

(2) C minimum size of the contact window;

(3) D minimum width of interconnection lines;

(4) DD minimum distance between the interconnection lines;

(5) CD minimum overlap of interconnection lines and contact windows;

(6) AB minimum distance between p-base and isolation groove.

The emitter length, L_E, is determined by the optimal emitter current per length as pointed out in Section 2.1.5 Equation (2.71)

$$L_E = I_E \frac{R_{SBi} d_B}{V_T \sqrt{2 B_N}} . \tag{2.140}$$

The emitter layout of high-frequency power transistors must be designed with a maximum possible emitter length-to-area ratio. An example is given in Fig. 2.28.

With the same vertical SBC structure p-n-p transistors can be produced (see Fig. 2.29(a), (b)) but the lateral p-n-p transistor in Fig. 2.29(b) has the disadvantage of small current gain (see Section 2.1.3) and the p-n-p substrate transistor in Fig. 2.29(a) has the disadvantage that all collectors are connected together.

A modern derivative of the standard SBC process is the ISOPLANAR process (Fig. 2.30). The side-wall isolation is made by SiO_2 grooves rather than reverse biased p-n junctions. This gives fewer parasitic effects (especially at high temperatures and under radiation) and makes self-aligned structures possible (see Fig. 2.2).

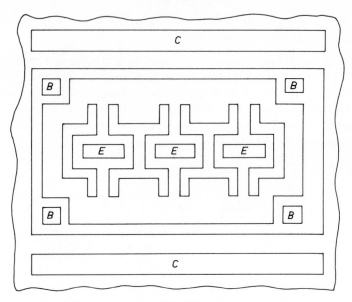

Fig. 2.28. Layout of a high-frequency power transistor.

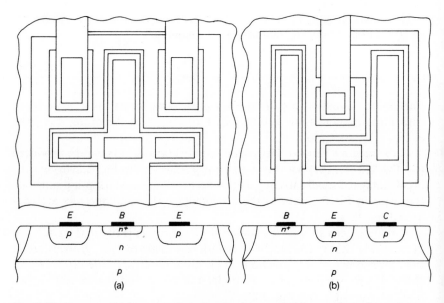

Fig. 2.29. Layout of (a) p-n-p substrate transistor and (b) p-n-p lateral transistor.

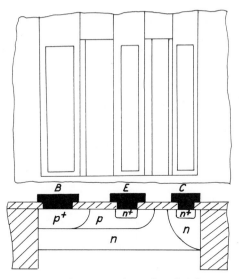

Fig. 2.30. Layout and cross-section of an Isoplanar transistor.

2.2 MOS FIELD-EFFECT TRANSISTORS

2.2.1 Principles of operation and MOS transistor structures for VLSI

The operation of a MOS transistor is based on the control of a conducting channel at or near the surface of a semiconductor with an insulated gate electrode. In Fig. 2.31 all four types of MOS field-effect transistors, together with their transfer characteristics, are compared.

1. The n-channel enhancement-type transistor in Fig. 2.31(a) is fabricated on a p-type substrate. The heavily doped n-regions are source (S) and drain (D). If we supply a voltage, V_{DS}, between drain and source the current, I_D, is nearly zero, but if we supply a high enough positive voltage, V_{GS}, at the gate a negative charge is created at the surface forming a conductive n-channel between source and drain (n) allowing an electron current. This drain current can be controlled by the gate voltage V_{GS}. However, not all of the negative charge is used for the enhancement of the n-channel; some is used to provide fixed charges such as ionized acceptors and the interface charge. Therefore a threshold voltage V_t exists (see Fig. 2.31(a), where $V_t \approx 1$ V).

2. The n-channel depletion-type transistor of Fig. 2.31(b) is fabricated by ion implantation of a shallow n-channel at the surface. Here a current flow is possible even at $V_{GS} = 0$ as can be seen from the transfer characteristic in Fig. 2.31(b). For negative gate voltages, $V_{GS} < 0$, the channel will be depleted of electrons and the current decreases. For a negative threshold

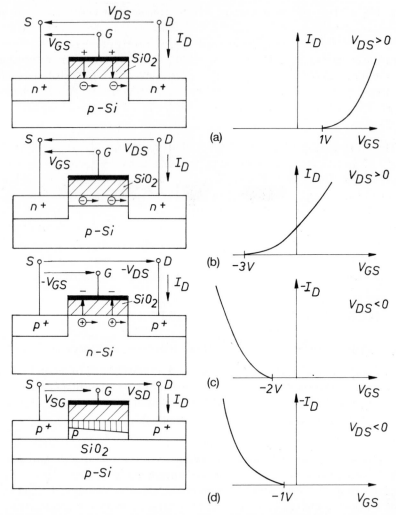

Fig. 2.31. Characteristics of different types of MOS transistors: (a) n-channel enhancement, (b) n-channel depletion, (c) p-channel enhancement, and (d) buried channel SOI transistor.

voltage, V_{tD} ($= -3$ V in this example) the current is zero. For $V_{GS} > 0$ the channel will be enhanced and the current increases.

3. If we start the process with an n-type substrate and use p-type source and drain regions we obtain p-channel enhancement-type transistors as shown in Fig. 2.31(c). For these transistors the signs of all voltages and currents are reversed but the operating principles are the same. n- and p-channel transistors are used together in CMOS circuits (see Chapter 3).

4. Fig. 2.31(d) shows a special MOS transistor structure, known as silicon-on-insulator (SOI), which is fabricated on an insulating substrate. Because of the SiO_2 insulation layer such transistors have very small parasitic capacitances and leakage currents and are therefore very promising for the future of VLSI. The operation of the device in Fig. 2.31(d) depends on deep depletion where a channel is enhanced under a surface space-charge layer. This will be explained in detail in Section 2.2.2 (Lim and Fossum 1983).

The cross-section of a typical CMOS structure for VLSI is shown in Fig. 2.32. n- and p-channel transistors are isolated by deep trenches preventing latch-up. The p-channel transistor is located in an n-well. Very shallow drain and source regions are used in VLSI (Lim and Fossum 1983; Miyake *et al.* 1989; Moravvey-Farshi and Green 1987).

In very high density VLSI and ULSI circuits the third (vertical) dimension must be used. As an example, a trench transistor cell for multimegabit memories is shown in Fig. 2.33. The MOS transistor is located in the vertical direction.

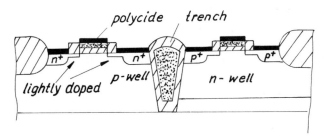

Fig. 2.32. CMOS structure for VLSI.

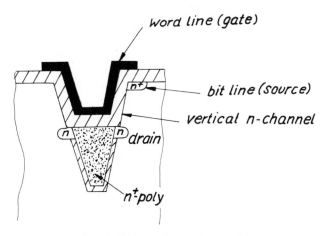

Fig. 2.33. Trench transistor cell.

2.2.2 I/V characteristics in strong and weak inversion

Enhancement transistors

For our analysis we use the n-channel model of Fig. 2.34. The drift current
in the n-channel is

$$I_D = enx_c\,\mu_n bE = \mu_n(-Q_n'')bE \tag{2.141}$$

where $enx_c = (-Q_n'')$ is the electron charge in the channel per unit area.
E is the electric field and μ_n is the electron mobility

$$E = \frac{\mathrm{d}V(y)}{\mathrm{d}y} \tag{2.142}$$

and $V(y)$ is the channel potential with respect to source ($V(0) = 0$,
$V(L) = V_{DS}$). Therefore the current becomes

$$I_D = \mu_n b(-Q_n'')\frac{\mathrm{d}V(y)}{\mathrm{d}y}. \tag{2.143}$$

The main problem here is to derive the channel charge ($-Q_n''$). The charge
neutrality condition at the MOS structure is

$$Q_n'' + Q_G'' + Q_B'' + Q_Z'' = 0 \tag{2.144}$$

where $Q_G'' > 0$ is the gate charge, $Q_Z'' > 0$ is the interface charge, and

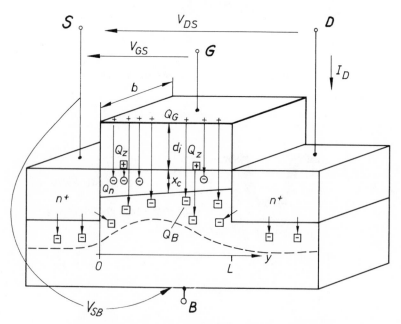

Fig. 2.34. Structure model of the MOS transistor.

$Q_B''< 0$ is the bulk charge of ionized acceptors, all per unit area.
With Equation (2.144) we have

$$(-Q_n'') = Q_G'' + Q_B'' + Q_Z'' . \qquad (2.145)$$

From Gauss's law we find the gate charge per unit area

$$Q_G'' = \frac{\epsilon_i}{d_i} V_i \qquad (2.146)$$

where V_i is the voltage drop on the gate insulator. We can find V_i with the
help of the energy-band diagram in Fig. 2.35.

Let us first consider the energy-band diagram for $V_G = 0$ (solid lines in
Fig. 2.35). In this case the Fermi level, W_F, is constant throughout the
structure. The total voltage drop is a built-in potential and is called the contact potential, V_K:

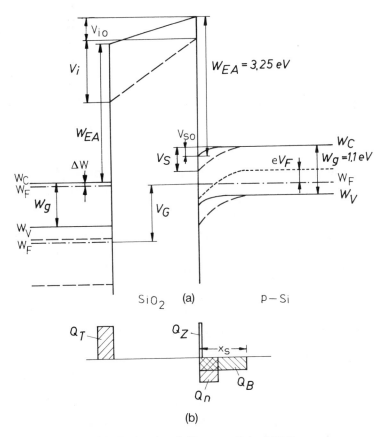

Fig. 2.35. Energy-band diagram of the MOS structure.

$$V_K = V_{io} + V_{so} \tag{2.147}$$

where V_{io} and V_{so} are the potential differences in the gate insulator and the semiconductor under thermal equilibrium, respectively. For heavily doped polysilicon gates (n^+) $\Delta W \approx 0$ and we find, from Fig. 2.35,

$$V_K = \frac{W_g}{2e} + V_F \tag{2.148}$$

where W_g is the energy-gap and V_F is given by

$$V_F = V_T \ln \frac{N_A}{n_i}. \tag{2.149}$$

If we supply a gate voltage V_G we have a difference in the Fermi levels of the gate and the semiconductor. The energy-band diagram for this case is shown in Fig. 2.35 with dashed lines. Now the gate voltage V_G is the sum of the excess voltage drops on the insulator and on the semiconductor. This gives

$$V_G = V_i + V_s - V_K \tag{2.150}$$

or

$$V_i = V_G - V_s + V_K \tag{2.151}$$

If we insert V_i of Equation (2.151) into Equation (2.146) the gate charge is

$$Q_G'' = \frac{\epsilon_i}{d_i} (V_G - V_s + V_K). \tag{2.152}$$

The gate voltage V_G is given by the gate–source voltage, V_{GS}, and the voltage drop, $V(y)$, along the channel (with respect to source)

$$V_G = V_{GS} - V(y).$$

The semiconductor surface potential V_s (the voltage drop on the semiconductor) is also dependent on the gate voltages as we have already shown in Fig. 1.24. To get simple analytical solutions we use two approximations shown in Fig. 2.36:

(1) *strong inversion* $V_{GS} > V_t$, $V_s \approx 2V_F$; $\tag{2.153}$

(2) *weak inversion* $V_{GS} < V_t$, $V_F < V_s < 2V_F$, $V_s \sim V_{GS}$. $\tag{2.154}$

Therefore the gate charge in strong inversion is

$$Q_G'' = \frac{\epsilon_i}{d_i} (V_{GS} - V(y) - 2V_F + V_K). \tag{2.155}$$

The bulk charge, Q_B'', due to negative ionized acceptors is

$$Q_B'' = -eN_A x_s \tag{2.156}$$

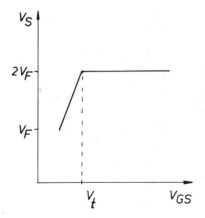

Fig. 2.36. Surface potential as a function of the gate voltage.

where x_s is the width of the space-charge layer at the semiconductor sur-face. It can be derived from Poisson's equation

$$x_s = \sqrt{\frac{2\epsilon_H V_s'}{eN_A}} \qquad (2.157)$$

and, with Equation (2.156),

$$Q_B'' = -\sqrt{2eN_A \epsilon_H V_s'} \ . \qquad (2.158)$$

The total potential difference, V_s', on the surface space-charge layer is (see Fig. 2.37)

$$V_s' = 2V_F + V_{SB} \qquad (2.159)$$

if a source–bulk voltage, V_{SB}, is supplied.

At the source, $y = 0$, we have $V(0) = 0$, and the bulk charge (per unit area) becomes

$$Q_B'' = -\sqrt{2eN_A \epsilon_H (2V_F + V_{SB})} \ . \qquad (2.160)$$

The interface state charge, Q_Z'', is considered to be constant here (e.g. $Q_Z'' = 10^{-8} \, \text{As/cm}^2$). With Q_B'' of Equation (2.160), Q_G'' of Equation (2.155) and Q_Z'' is constant, we find the electron charge in the channel

$$(-Q_n'') = \frac{\epsilon_i}{d_i} \left\{ V_{GS} - V(y) - \left(2V_F + V_{FB} + \frac{d_i}{\epsilon_i} \sqrt{2eN_A \epsilon_H (2V_F + V_{SB})} \right) \right\} \qquad (2.161)$$

or

$$(-Q_n'') = \frac{\epsilon_i}{d_i} \left(V_{GS} - V(y) - V_t \right) \qquad (2.162)$$

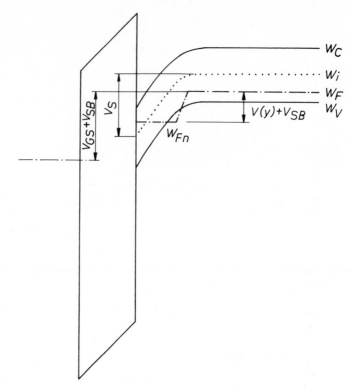

Fig. 2.37. Energy-band diagram of the MOS structure with substrate bias.

where we have defined a threshold voltage V_t. If we compare Equation (2.162) with Equation (2.161) we find

$$V_t = 2V_F + V_{FB} + \frac{d_i}{\epsilon_i}\sqrt{2eN_A\epsilon_H(2V_F + V_{SB})} \qquad (2.163)$$

where V_{FB} is the flat-band voltage (see also Section 1.3)

$$V_{FB} = \left(-V_K - \frac{d_i}{\epsilon_i}Q_Z''\right). \qquad (2.164)$$

The threshold voltage V_t in Equation (2.163) can be interpreted as the gate–source voltage V_{GS} at which the channel disappears. For $V_{GS} > V_t$ a strong inversion channel starting at the source will be built up. The linear rise $Q_n''(0) \sim (V_{GS} - V_t)$ is shown in Fig. 2.38.

Integrating the current flow equation

$$I_D = \mu_n b(-Q_n'')\frac{dV}{dy} = \mu_n b\frac{\epsilon_i}{d_i}(V_{GS} - V(y) - V_t)\frac{dV}{dy} \qquad (2.165)$$

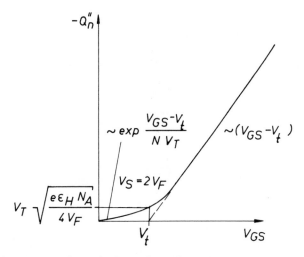

Fig. 2.38. Electron charge in the n-channel as a function of gate voltage.

from $y = 0$ ($V(0) = 0$) to $y = L$ ($V(L) = V_{DS}$, see Fig. 2.34) yields

$$I_D = \mu_n \frac{\epsilon_i}{d_i} \frac{b}{L} \left\{ (V_{GS} - V_t) V_{DS} - \frac{V_{DS}^2}{2} \right\}. \qquad (2.166)$$

This is plotted in Fig. 2.39. But Equation (2.166) is valid only for $V_{DS} \le V_{GS} - V_t$ (up to the dashed line in Fig. 2.39). For $V_{DS} > V_{GS} - V_t$ the

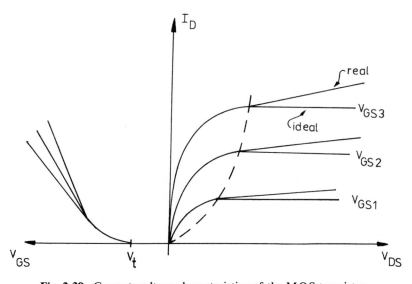

Fig. 2.39. Current–voltage characteristics of the MOS transistor.

channel charge, Q_n'', at the drain end of the channel is zero (no channel). Therefore the integration of Equation (2.165) along the whole channel $(0 \leq y \leq L)$ to get the current–voltage characteristic is prohibited. For $V_{DS} > V_{GS} - V_t$ we call the operation *pinch-off* mode.

A very simple current flow model for the pinch-off region is derived by setting $V_{DS} = V_{GS} - V_t$ in Equation (2.166):

$$I_D = \tfrac{1}{2}\mu_n \frac{\epsilon_i}{d_i}\frac{b}{L}(V_{GS} - V_t)^2. \qquad (2.167)$$

In this case the drain current in the pinch-off region is constant, independent of the drain voltage V_{DS} but for real transistors there is a finite rise of the output characteristics (see Fig. 2.39) in the pinch-off region. This can be qualitatively explained as follows. In Equation (2.167) the channel length L is not the constant distance between drain and source but the electronic channel length which decreases in pinch-off with increasing drain voltage V_{DS}. This gives an increase of the drain current as shown in Fig. 2.39.

A detailed analysis of the I/V characteristics in the pinch-off region is rather complex (Murphy 1980) and only possible with numerical simulations. Therefore we adopt an empirical model which gives, with the model parameter K_4:

$$I_D = \tfrac{1}{2}\mu_n \frac{\epsilon_i}{d_i}\frac{b}{L}(V_{GS} - V_t)\{(V_{GS} - V_t) + K_4 V_{DS}\}. \qquad (2.168)$$

To prevent discontinuities in the I/V characteristics at the pinch-off point, $V_{DS} = V_{GS} - V_t$, we have also to correct Equation (2.166) slightly and write

$$I_D = \tfrac{1}{2}\mu_n \frac{\epsilon_i}{d_i}\frac{b}{L}\left\{(V_{GS} - V_t)(2 + K_4) - V_{DS}\right\}V_{DS}. \qquad (2.169)$$

The transfer characteristic $I_D = f(V_{GS})$ is also shown in Fig. 2.39.

For $V_{GS} < V_t$ we have $I_D = 0$ in the strong inversion theory. But this is not true if we also consider very small currents (in the nanoampere and picoampere range). For $V_{GS} < V_t$ and $V_s > V_F$ a weak inversion channel exists at the surface (see the energy-band diagram in Fig. 2.35). In this channel the electron charge, $-Q_n''$, per unit area is (Fig. 2.38, (Möschwitzer and Lunze 1990))

$$(-Q_n'') = V_T\sqrt{\frac{e\epsilon_H N_A}{2(2V_F + V_{SB})}}\exp\frac{V_{GS} - V_t - V(y)}{NV_T} \qquad (2.170)$$

where

$$N = 1 + \frac{d_i}{\epsilon_i}\sqrt{\frac{eN_A\epsilon_H}{2(2V_F + V_{SB})}} \approx 1. \qquad (2.171)$$

The integration of

$$I_D = \mu_n b(-Q_n'')\frac{dV}{dy} =$$

$$\mu_n bV_T \sqrt{\frac{e\epsilon_H N_A}{2(2V_F + V_{SB})}} \exp\frac{V_{GS} - V_t - V(y)}{NV_T}\frac{dV}{dy} \quad (2.172)$$

from $y = 0$ ($V(0) = 0$) to $y = L$ ($V(L) = V_{DS}$) yields the drain current in weak inversion

$$I_D = I_o\left(\exp\frac{V_{GS} - V_t}{NV_T}\right)\left\{1 - \exp\left(-\frac{V_{DS}}{V_T}\right)\right\} \quad (2.173)$$

where

$$I_o = \mu_n\frac{b}{L}V_T^2\sqrt{\frac{e\epsilon_H N_A}{2(2V_F + V_{SB})}} . \quad (2.174)$$

In Fig. 2.40 the transfer characteristics in strong and weak inversion are depicted.

Up to now we have considered the n-channel enhancement-mode transistor. Exactly the same results are obtained for the p-channel enhancement-mode transistor if we use $-I_D$, V_{SG}, and V_{SD} in Equations (2.166)–(2.169) and (2.173), instead of I_D, V_{GS}, and V_{DS}. We find

$$V_{tp} = 2V_F - V_{FB} + \frac{d_i}{\epsilon_i}\sqrt{2eN_D\epsilon_H(2V_F + V_{BS})} . \quad (2.175)$$

For n- and p-channel enhancement-mode transistors the threshold voltage

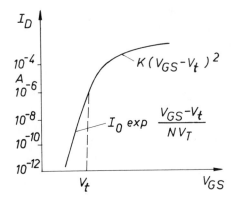

Fig. 2.40. Current–voltage characteristic of the MOS transistor including weak inversion current.

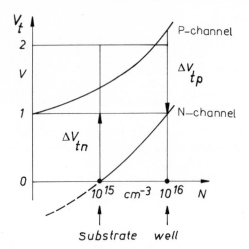

Fig. 2.41. The dependence of the threshold voltage of n- and p-channel MOS transistors on the substrate doping concentration.

increases with increasing substrate doping N_A and N_D (see Fig. 2.41). The difference in the p- and n-channel threshold voltage is $2\,V_{FB}$ in this example. It can be calculated that for low substrate dopings (e.g. $N_D \approx 1 \times 10^{15}\,\text{cm}^{-3}$) and zero source–bulk voltage ($V_{SB} = 0$) the n-channel threshold voltage is very small ($V_t \approx 0$). If we want useful threshold voltages for n-channel enhancement-mode transistors in the range 0.5–1 V we can supply a certain source–bulk voltage or increase the bulk charge by ion implantation of acceptors (e.g. boron).

For CMOS circuits the p- and n-channel threshold voltages should be the same. (e.g. ≈ 0.5 V) and this can be accomplished by ion implantation of boron (acceptor) atoms. It yields a negative shift $-\Delta V_{tp}$ of the p-channel threshold voltage and a positive shift $+\Delta V_{tn}$ of the n-channel threshold voltage (see Fig. 2.41)

$$-\Delta V_{tp} = +\Delta V_{tn} = e\,\frac{d_i}{\epsilon_i}\,D_B'' \qquad (2.176)$$

where D_B'' is the boron dose.

We will demonstrate this threshold voltage adjustment for the n-well CMOS structure of Fig. 2.32 for which we have the following numerical example: acceptor concentration of the epitaxy layer, $N_A = 10^{15}\,\text{cm}^{-3}$; donor concentration of the n-well, $N_D = 10^{16}\,\text{cm}^{-3}$; and $\epsilon_i/d_i = 1.6 \times 10^{-7}$ As/Vcm². For this example Fig. 2.41 yields a required threshold shift of approximately $\Delta V_t \approx 1$ V. The required boron dose for this example is therefore

$$D_B'' = \frac{\epsilon_i V_t}{e d_i} = \frac{1.6 \times 10^{-7}\,\text{As}\,1\,\text{V}}{1.6 \times 10^{-19}\,\text{As cm}^2\,\text{V}} = 10^{12}\,\text{cm}^{-2}. \qquad (2.177)$$

Depletion transistors

Depletion transistors are made by implantation of a conducting channel d_K, at the surface of the semiconductor substrate (Fig. 2.42) or they can arise because of a positive interface charge, $Q_Z > 0$. This conducting channel, may be depleted by a negative gate voltage ($V_{GS} < 0$ for n-channel) from the top (depletion layer width x_s) and by a substrate bias ($V_{SB} > 0$ for n-channel) from the bottom (depletion layer width d_s) so that the conducting channel is buried. We mention, without proof, that these transistors can be modelled by the same I/V characteristics Equations (2.166)–(2.169) if we use, instead of a positive threshold voltage V_t, a negative threshold voltage $V_{tD} < 0$. This depletion threshold voltage will now be derived. We follow the same reasoning as we have already done for the enhancement threshold voltage.

From the charge neutrality equation we have

$$-Q_n'' = Q_G'' + Q_B'' + Q_Z'' \qquad (2.178)$$

and the gate charge per unit area is (see Equation (2.152))

$$Q_G'' = \frac{\epsilon_i}{d_i} (V_{GS} - V(y) - V_s + V_K). \qquad (2.179)$$

The semiconductor surface potential V_s (the voltage drop on the surface depletion region x_s) can be calculated from Poisson's equation with the constant space charge $\rho = eN$. We find

$$V_s = \frac{-eNx_s^2}{2\epsilon_H} \qquad (2.180)$$

where N is the donor concentration in the channel.

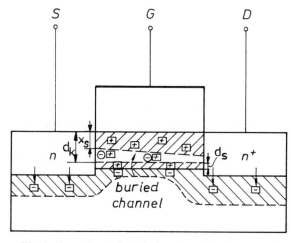

Fig. 2.42. n-channel depletion-type MOS transistor.

The bulk charge per unit area is

$$Q_B'' = eNx_s. \tag{2.181}$$

The space-charge layer width, d_s, of the channel–substrate p-n junction is (see Section 1.4)

$$d_s = \sqrt{\frac{2\epsilon_H N_A}{eN(N + N_A)}} \, (V_{SB} + V_{bi}) \tag{2.182}$$

where V_{bi} is the built-in voltage of the channel–substrate p-n junction, ϵ_H is the permittivity of the semiconductor, and N_A is the impurity concentration (acceptors) in the substrate. With V_s, Q_B'', and the flat-band voltage, V_{FB} (Equation (2.164)), we find, from Equation (2.178),

$$-Q_n'' = \frac{\epsilon_i}{d_i}\left(V_{GS} - V(y) + \frac{eNx_s^2}{2\epsilon_H} - V_{FB} + \frac{d_i}{\epsilon_i} eNx_s\right). \tag{2.183}$$

We define the depletion threshold voltage V_{tD} as the gate–source voltage where $-Q_n''(0) = 0$. This yields

$$V_{tD} = -\left(\frac{eNx_s^2}{2\epsilon_H} - V_{FB} + \frac{d_i}{\epsilon_i} eNx_s\right). \tag{2.184}$$

For $Q_n''(0) = 0$, x_s is given by $x_s = d_K - d_s$, using d_s from Equation (2.182). $-V_{tD}$ increases with increasing channel doping and decreasing source–bulk voltage V_{SB}.

SOI transistors

Silicon-on-insulators (SOI) transistors are fabricated on an insulating substrate rather than in a semiconductor bulk. This gives better isolation of the individual devices and smaller parasitic elements (Lim and Fossum 1983; McKitterich and Caviglia 1989; Veeraraghavan and Fossum 1988; Young 1989). The channel (of thickness d_K) is isolated by a SiO$_2$ layer from the substrate as is shown in Fig. 2.43. The channel charge can be controlled by the normal (top) gate and also the substrate (bottom) gate. If the thin channel is lightly doped with acceptors (of concentration N_A) an n-channel can be enhanced at the surface by a positive gate–source voltage $V_{GS} > 0$. The energy-band diagram for this case is shown in Fig. 2.44(a). We assume that the channel layer is thick enough so that $x_s + d_s < d_K$. This means the space-charge layers at the top (0) and at the bottom (d_K) do not touch one another. The potential of the neutral p-layer is floating (Veeraraghavan and Fossum 1988) and can be roughly estimated as

$$V_f = V_S + 2V_T \ln \frac{I_R}{I_S} \approx V_S \tag{2.185}$$

where V_S is the source potential. $I_R \approx I_S$ is the reverse current of the

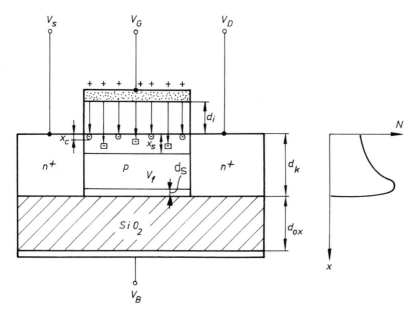

Fig. 2.43. SOI transistor.

drain–substrate diode. The enhancement threshold voltage is therefore (with $V_S = 0$)

$$V_{tE} \approx 2V_F + V_{FB} + \frac{d_i}{\epsilon_i} \sqrt{2eN_A \epsilon_H (2V_F)} . \qquad (2.186)$$

If the channel layer is doped with donors (of concentration N_D) we get depletion-mode transistors with buried channels as we have discussed in Section 2.2.2.

If we use very thin n-layers so that at $V_{GS} = 0$ the channel layer is fully depleted ($x_s = d_k$) we get buried channel deep depletion-mode transistors. The energy-band diagram for this case is sketched in Fig. 2.44(b).

The semiconductor surface potential under the gate $V_s = \varphi(0)$ is, at a fully depleted n-layer for $d_{ox} \gg d_i$, $V_{SB} = 0$, and the boundary condition $\varphi(d_k) = 0$,

$$V_s = -eN_D \frac{d_k^2}{2\epsilon_H} \qquad (2.187)$$

and the bulk charge per unit area is

$$Q_B'' = eN_D d_k . \qquad (2.188)$$

A simplified expression for the threshold voltage of this deep-depletion transistor is, for $V_{SB} = 0$ and $d_{ox} \gg d_i$ (neglecting the influence of the substrate)

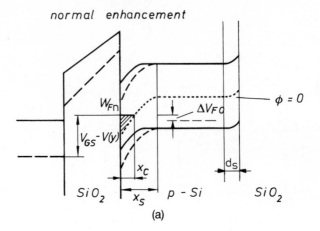

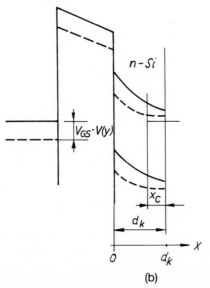

Fig. 2.44. Energy-band diagram of (a) normal enhancement-type and (b) deep depletion-type SOI transistor.

$$V_{tDD} = V_{FB} - eN_D d_k \left(\frac{d_k}{2\epsilon_H} + \frac{d_i}{\epsilon_i} \right). \tag{2.189}$$

It is positive if

$$V_{FB} > eN_D d_k \left(\frac{d_k}{2\epsilon_H} + \frac{d_i}{\epsilon_i} \right) \tag{2.190}$$

and we have a deep-depletion transistor with buried channel. If we use very thin fully depleted p-layers we get enhancement-type transistors with a threshold voltage $v_{tDE} < 0$.

Fully depleted SOI MOSFETs have smaller floating-substrate and short-channel effects and higher transconductances than others (McKitterich and Caviglia 1989; Young 1989). The flat-band voltage, V_{FB}, which is responsible for the sign of the threshold voltage can be altered by the doping of the polysilicon gate (see Equation (2.164) and Fig. 2.35).

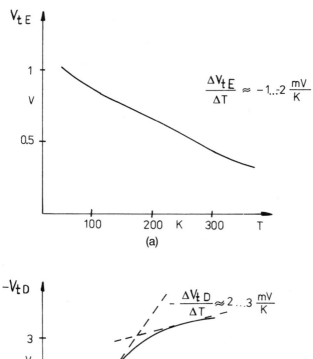

Fig. 2.45. Temperature dependence of the threshold voltage of (a) enhancement-type and (b) depletion-type transistor.

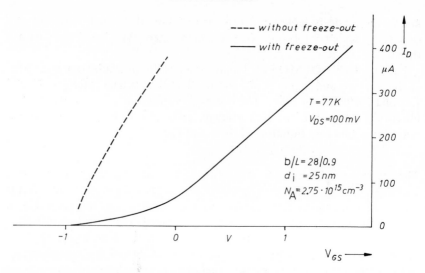

Fig. 2.46. The influence of the impurity freeze-out on the current–voltage characteristic of MOS transistors.

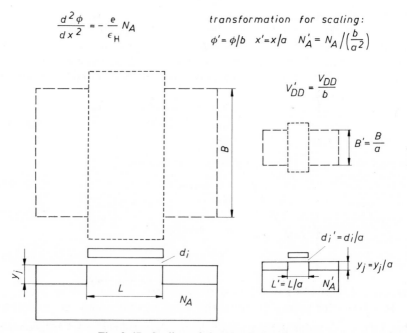

Fig. 2.47. Scaling of the MOS transistor.

2.2.3 Temperature effects

MOS transistor currents are much less dependent on temperature than the currents in bipolar transistors (see Section 2.1.5). The temperature dependence is mainly due to the temperature dependence of the threshold voltage (Selberherr 1989). For enhancement-type transistors we have approximately (Fig. 2.45)

$$\frac{dV_t}{dT} \approx \frac{dV_F}{dT} \approx -1 \, mVK^{-1}. \tag{2.191}$$

The temperature dependence of the leakage currents (e.g. the weak inversion current) will be considered in Section 2.2.4. A special temperature effect is due to the impurity freeze-out. At very low temperatures the impurities in the semiconductor bulk are not completely ionized. This results in a flattening of the I/V characteristic as shown in Fig. 2.46. With increasing gate voltage the electron density in the channel increases and the ionization rate of the impurities decreases. On the other hand if we decrease the gate voltage more and more impurities will be ionized and the channel does not disappear abruptly at a threshold voltage (see Fig. 2.46) (Simoen $et\,al.$ 1989).

2.2.4 Small-geometry effects

Device scaling
As the technology advances single devices in integrated circuits are scaled to smaller and smaller dimensions in order to place more devices on a chip and to improve their electronic behaviour, mainly their operating speed. Ideal scaling means scaling of all dimensions x, potentials φ, and doping concentrations N_A, so that Poisson's equation

$$\frac{d^2\varphi}{dx^2} = -\frac{e}{\epsilon_H} N_A \tag{2.192}$$

is kept unchanged. This is true if we scale (see Fig. 2.47)

$$\varphi' = \varphi/b \tag{2.193}$$

$$x' = x/a \tag{2.194}$$

$$N_A' = N_A \Big/ \left(\frac{b}{a^2}\right) \tag{2.195}$$

with $a > 1$, $b > 1$. Table 2.1 shows an example of such scaling.

However, for many reasons the ideal scaling rules of Equations (2.193)–(2.195) cannot be accomplished exactly. The parameters which have not been scaled in the past are the supply voltage, V_{DD} (because of TTL

Table 2.1. Example for device scaling

	$a = 1, b = 1$	$a = 5, b = 2.5$
V_{DD} (V)	5	2
N_A (cm^{-3})	3×10^{15}	3×10^{16}
d_i (nm)	60	12
L (μm)	2.5	0.5
y_j (μm)	0.6	0.12

compatibility), and the drain/source junction depth, y_j (because of process constraints). The parameters which cannot be scaled are built-in potentials and leakage currents (e.g. the weak inversion current). These cause unwanted effects in the device behaviour (El Mansy 1982; Lee et al. 1988):

(1) short- and narrow-channel effects on the threshold voltage (Akers and Sanches 1982; Akers et al. 1987; Wang 1987; Yau 1975);

(2) increase of subthreshold currents (weak inversion current and punch-through current) (Chatterjee et al. 1979; Punbley and Meindl 1989; Troutman 1979; Wu et al. 1985);

(3) high-field effects (El-Banna and El-Nokali 1989; Cotrell et al. 1979; Troutman et al. 1980; Wong et al. 1987; Hwang and Dutton 1989)

(4) parasitic transistor and thyristor effects (Menozzi et al. 1988; Yang and Wu 1989).

These will be considered in detail now.

Short- and narrow-channel effects
The decrease of the threshold voltage at small channel length is depicted in Fig. 2.48. This short-channel effect is due to charge sharing of the bulk

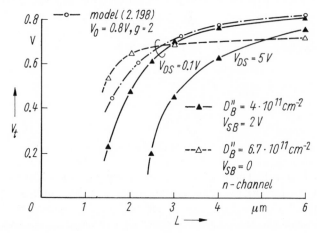

Fig. 2.48. The dependence of the threshold voltage on the channel length.

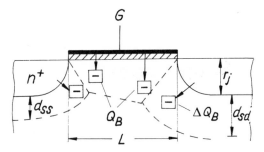

Fig. 2.49. Charge sharing at the short-channel MOS transistor.

charge Q_B as shown in Fig. 2.49 (Akers and Sanches 1982; Yau 1975). For small channel lengths the ratio of bulk charge shared with the gate to bulk charge shared with drain and source space-charge layers decreases. With simple geometrical considerations (see Fig. 2.49) we obtain the dependence of the threshold voltage, V_t on the channel length L (Yau 1975)

$$V_t = 2V_F + V_{FB} + \left(\frac{d_i}{\epsilon_i}\sqrt{2\,\epsilon_H e N_A (V_{SB} + 2V_F)}\right)F_1(L) \qquad (2.196)$$

with

$$F_1(L) = \left\{1 - \left(\sqrt{1 + \frac{2d_s}{r_j}} - 1\right)\frac{r_j}{L}\right\} \qquad (2.197)$$

where, in Equation (2.196), we assumed equal space-charge layer widths of the source–substrate and drain–substrate p–n junctions.

Although this simple model is straightforward and easily to understand, Equation (2.196) agrees only qualitatively with measured data. Several attempts have been made to improve this simple model (Akers and Sanches 1982) but a better way is to calculate $V_t(L)$ by numerical simulations (Cham 1986; Selberherr 1984) and adopt, for analytical calculations, an empirical model such as

$$V_t = V_{t\infty} - \frac{V_o}{(L/\mu m)^g} \qquad (2.198)$$

with $g > 1$ and V_o, by curve fitting.

If we decrease the channel width b an increase or decrease of the threshold voltage depending on the side-wall oxide isolation can occur (Akers and Sanches 1982; Akers et al. 1987; Wang 1987). For non-recessed oxide (such as the bird's beak of Fig. 2.50) we have the normal narrow-width effect. Here the threshold voltage increases with decreasing channel width b due to the fringing field at the taper region of the gate and the field oxide.

In this case more bulk charge Q_B is shared with the gate so that the threshold voltage is increased.

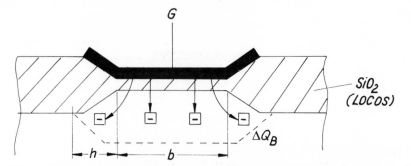

Fig. 2.50. Fringing fields at the taper region of a narrow-channel transistor.

The same simple geometrical model used for the short-channel effect yields

$$V_t = 2V_F + V_{FB} + \left(\frac{d_i}{\epsilon_i}\sqrt{2\,\epsilon_H eN_A(V_{SB} + 2V_F)}\right)F_2(b, L) \quad (2.199)$$

where

$$F_2(b, L) = \left[1 + \frac{h}{b}\left\{1 + \frac{2}{3}\frac{r_j}{L}\left(\sqrt{1 + \frac{2d_s}{r_j}} - 1\right)\right\}\right]. \quad (2.200)$$

In the case of deep trench isolation (Fig. 2.51) we have a decrease of the threshold voltage with decreasing channel width because of lateral fields. This effect is called the inverse narrow-width effect (Akers *et al.* 1987).

The lateral field increases the channel charge Q_n'' and we have (Akers *et al.* 1987)

$$V_t = 2V_F + V_{FB} + \left(\frac{d_i}{\epsilon_i}\sqrt{2\epsilon_H eN_A(V_{SB} + 2V_F)}\right)F_3(b, L) \quad (2.201)$$

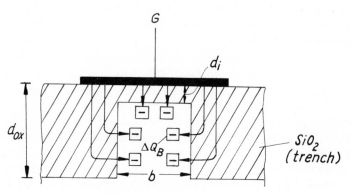

Fig. 2.51. Fringing fields at the boundaries of a narrow-channel transistor with trench isolation.

where

$$F_3 = \left[1 - \left(\sqrt{1 + \frac{2d_s}{r_j}} - 1\right)\right] \frac{r_j}{L} \left(1 + \frac{4d_i}{\pi b} \ln 2 \frac{d_{ox}}{d_i}\right)^{-1} \quad (2.202)$$

and

$$d_s = \sqrt{\frac{2 \epsilon_H}{e N_A} (2V_F + V_{SB})}. \quad (2.203)$$

These models are adequate for a qualitative explanation of the effect but for useful quantitative results two- or three- dimensional computer simulations must be done.

Subthreshold currents

For gate voltages smaller than the threshold voltage we still have very small subthreshold currents. These are

(1) weak inversion current;

(2) punch-through current;

(3) reverse-currents of the p-n junctions.

We have already considered the weak inversion current in Section 2.2.2. With Equation (2.173) we obtain for $V_{DS} \gg V_T$, $V_{GS} = 0$

$$I_{DW} = I_o \exp - \frac{V_t}{V_T}. \quad (2.204)$$

In Fig. 2.52 some results of computer simulations are plotted. The increase of the weak inversion current with decreasing channel length L is mainly due to the short-channel effect (decreasing threshold voltage, see Fig. 2.48). On the other hand for ideal scaling the current should decrease as the device dimensions are scaled down. Unfortunately the results in Fig. 2.52 show the opposite effect.

The temperature dependence of the weak inversion current is shown in Fig. 2.53. This is mainly due to the temperature dependence $V_T = kT/e$ in the exponent of Equation (2.204) (Selberherr 1989). At small channel lengths, L, and high drain voltages, the space-charge layers of the drain–substrate and the source–substrate p-n junction come into contact (Fig. 2.54). This effect is called punch-through and the potential distribution is plotted in Fig. 2.55. There is a potential barrier V_B with a saddle point at y_p. The punch-through current flows underneath the surface across this saddle point in the potential hill (indicated by an arrow in Fig. 2.55). The punch-through current is a barrier-limited emission current (see Section 1.5.2, Equation (1.112)

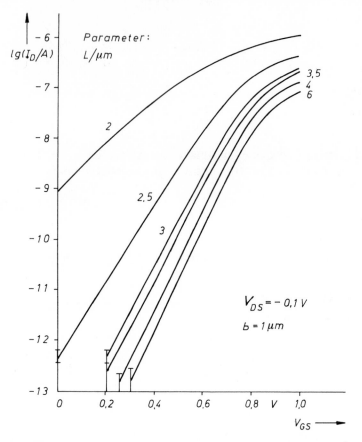

Fig. 2.52. Current–voltage characteristics for different channel lengths.

$$I_{DP} = I_{po} \exp - \frac{V_B}{V_T} \qquad (2.205)$$

where

$$I_{po} = eA \frac{N_D v_{th}}{4} \qquad (2.206)$$

and V_B is the barrier height, N_D is the source doping concentration, v_{th} is the thermal velocity, and A is the channel area.

The barrier height can be lowered by the drain–source voltage $V_B = V_{BO} - V_{DS}/V_O$ and increased by the source–bulk voltage V_{SB} (Troutman 1979). Therefore we have an exponential dependence of the punch-through current on the drain–source voltage. This is verified by the results of computer simulations in Fig. 2.56. We can also see from Fig. 2.56

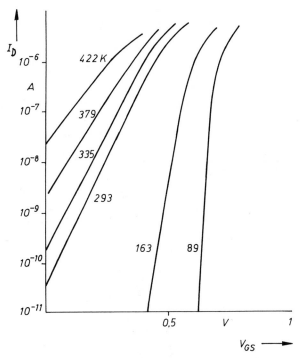

Fig. 2.53. Current–voltage characteristics for different temperatures.

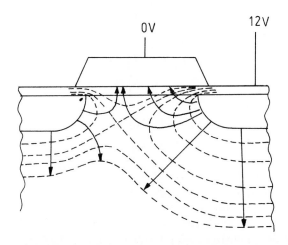

Fig. 2.54. Electric field and potential distribution at punch-through.

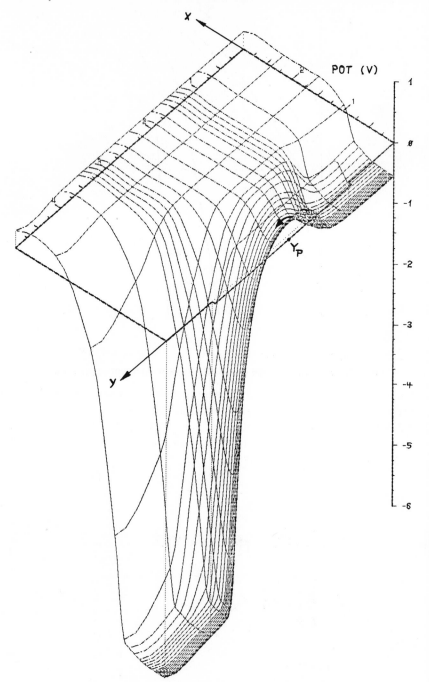

Fig. 2.55. Two-dimensional potential distribution at punch-through showing the current flow across a saddle in the potential hill.

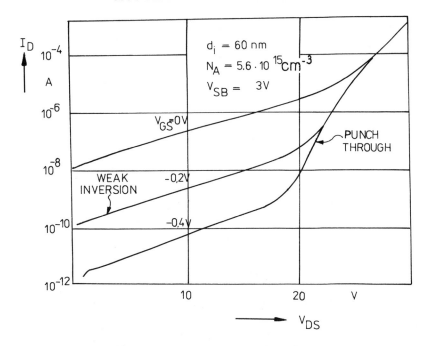

Fig. 2.56. Punch-through and weak inversion current.

that at low drain voltages the subthreshold current is only weakly dependent on the drain voltage. This is why at small drain voltages the subthreshold current is mainly a weak inversion current rather than a punch-through current.

In Fig. 2.57 we show the current distribution underneath the semiconductor surface for different implantation profiles of boron atoms. In the case

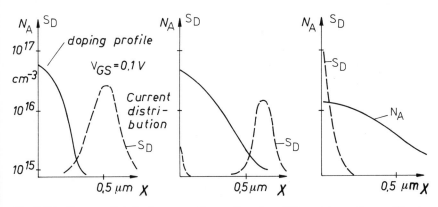

Fig. 2.57. Current distribution near the surface for different impurity profiles in MOS transistors.

of flat profiles with low boron concentrations the subthreshold current is mainly a weak inversion current at the surface. For steep profiles with high concentrations at the surface the subthreshold current is mainly a punch-through current in a buried channel underneath the surface.

The reverse currents of the p-n junctions are (see Section 1.4.4) (Chatterjee *et al.* 1979)

$$I_R = eA \left(\frac{n_i d_s}{\tau_s} + D_n \frac{n_i^2}{N_A L_n} \right) \qquad (2.207)$$

where d_s is the space-charge layer width, τ_s is the carrier lifetime, N_A is the substrate doping concentration, D_n is the diffusion coefficient, and L_n is the diffusion length of electrons in the substrate. We then have

$$d_s = \sqrt{\frac{2\,\epsilon_H}{e\,N_A}\,(V_{bi} + V_{DS} + V_{SB})}\,. \qquad (2.208)$$

At low temperatures the first part of Equation (2.207) dominates (generation current in the space charge layer $\sim n_i$) and at high temperatures the second part dominates (diffusion current $\sim n_i^2$). An example is shown in Fig. 2.58.

High-field effects

At high electric fields in the gate oxide and at the semiconductor surface we have lower carrier mobilities at the surface, emission and tunnel currents into

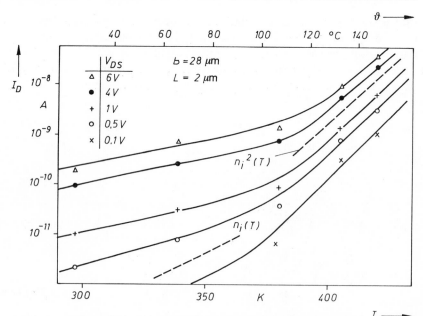

Fig. 2.58. Temperature dependence of leakage currents.

the oxide, and breakdown effects in the oxide and space-charge layers of the p-n junctions (Cotrell *et al.* 1979; El-Banna and El-Nokali 1989; Troutman *et al.* 1980; Wong *et al.* 1987).

A useful formula for the field dependence of the mobility is

$$\mu(E) = \frac{\mu_{\mathrm{o}}}{1 + \dfrac{\mu_{\mathrm{o}}E}{v_{\mathrm{g}}}} \tag{2.209}$$

where μ_0 is the low-field mobility and v_{g} is the saturation velocity (see Fig. 1.12).

The electric field E can be composed of an x and a y component:

$$E = \sqrt{(a_x E_x)^2 + (a_y E_y)^2}. \tag{2.210}$$

With the parameters $\mu_{\mathrm{o}} = 1{,}200\ \mathrm{cm^2/Vs}$, $v_{\mathrm{g}} = 1.2 \times 10^7\ \mathrm{cm/s}$, $a_x = 0.85$, and $a_y = 0.08$ for n-channel transistors, and $\mu_{\mathrm{o}} = 500\ \mathrm{cm^2/Vs}$, $v_{\mathrm{g}} = 9 \times 10^6\ \mathrm{cm/s}$, $a_x = 0.9$, and $a_y = 0.65$ for p-channel transistors we obtained good agreement with experimental data (see Figs 2.59 and 2.60).

At high drain voltages we have field peaks and peaks of the avalanche generation rate G at the drain edge as is plotted in Fig. 2.61. Using the analytic approximation for the avalanche generation rate

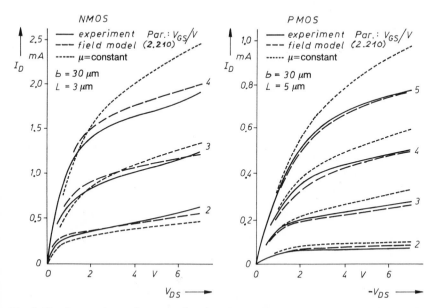

Fig. 2.59. Comparison of measured and simulated current–voltage characteristics of an n-channel transistor.

Fig. 2.60 Comparison of measured and simulated current–voltage characteristics of a p-channel transistor.

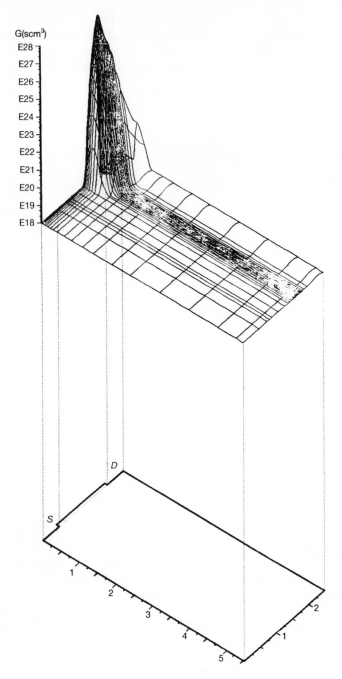

Fig. 2.61. Avalanche ionization rate has a maximum near the drain ($V_{GS} = 1$ V, $V_{DS} = 7$ V, $V_{BS} = 0$).

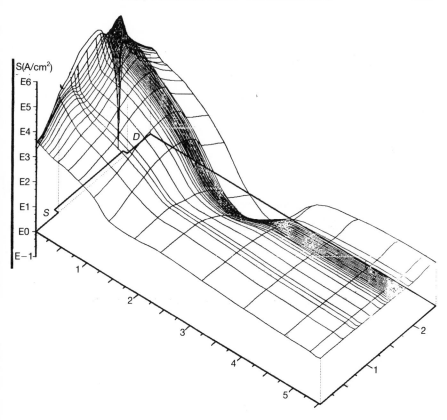

Fig. 2.62. Current distribution at avalanche breakdown ($V_{GS} = 1$ V, $V_{DS} = 7$ V, $V_{BS} = 0$).

$$G = \frac{|S_n|}{e} A_n \exp\left(-\frac{B_n}{E_y}\right) + \frac{|S_p|}{e} A_p \exp\left(-\frac{B_p}{E_y}\right) \qquad (2.211)$$

with $A_n = 7 \times 10^5\,\text{cm}^{-1}$, $B_n = 1.2 \times 10^6\,\text{V/cm}$, $A_p = 1.5 \times 10^6\,\text{cm}^{-1}$, and $B_p = 2 \times 10^6\,\text{V/cm}$ we obtain an avalanche current distribution near the drain as shown in Fig. 2.62. In Fig. 2.63 measured and simulated I/V characteristics including the avalanche effect are compared.

At high fields in the pinch-off region and in the gate oxide electrons can be emitted via the Si–SiO$_2$ potential barrier and move into the gate oxide. Because of these hot electron emissions we have a gate current, normally in the picoampere range. Figure 2.64 shows simulated gate currents employing a lucky electron model. At $V_{GS} \approx V_{DS}$ the gate current has a maximum. The substrate current due to the avalanche effect is also shown.

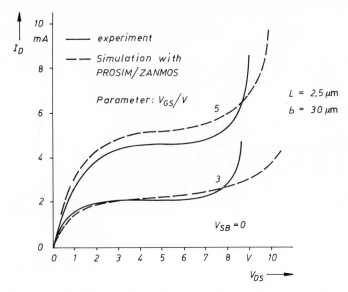

Fig. 2.63. Current–voltage characteristics with avalanche breakdown.

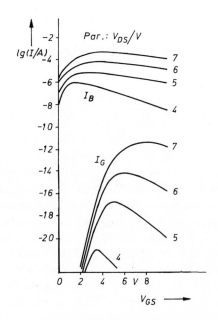

Fig. 2.64. Hot electron gate and substrate currents.

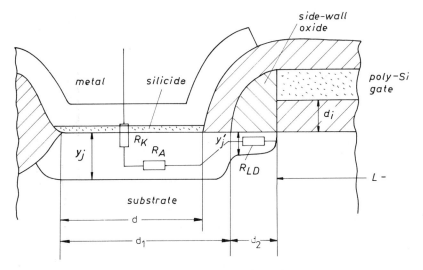

Fig. 2.65. Drain/source series resistors.

Effect of series resistors

Small transistor geometries cause high spreading resistors. Some details of source/drain spreading resistors are shown in Fig. 2.65. These consist of three parts:

$$R_{SS'} = R_K + R_A + R_{LD}.$$ (2.212)

The contact resistance, R_K, with the results of Sections 1.2 and 1.5.1, and Fig. 1.19,

$$R_K = \frac{1}{b}\sqrt{\rho_K R_S} \, \coth\sqrt{\frac{R_S}{\rho_K}} \, d.$$ (2.213)

In Figs 2.66 and 2.67 the dependence of the specific contact resistivity, ρ_K, and the sheet resistance, R_s, on the width of the contact window, d, and the source/drain depth, y_j, is shown. It should be mentioned that ρ_K and R_S increase very rapidly with decreasing d and y_j.

The spreading resistances of the deep source/drain and the lightly doped source/drain are

$$R_A = R_{SA}\frac{d_1}{b}$$ (2.214)

and

$$R_{LD} = R_{SL}\frac{d_2}{b},$$ (2.215)

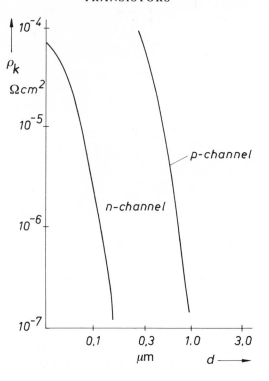

Fig. 2.66. Specific contact resistivity.

respectively. R_{SA} and R_{SL} can be obtained from Fig. 2.67 for every y_j. Finally the dependence of the mobility on the channel length is shown in Fig. 2.68.

Parasitic transistor and thyristor effects (snap-back, latch-up)

Under avalanche breakdown conditions we have an increased substrate current, I_B. Because of these substrate currents a substrate voltage drop $\Delta V = I_B R_S$ can bias the source–substrate p-n junction in the forward direction and switch the source–substrate–drain bipolar transistor (n-p-n) on. This bipolar transistor action causes a snap-back of the drain–source voltage, V_{DS}.

A parasitic thyristor effect, called latch-up, can be observed in CMOS structures. The n-p-n-p structure in Fig. 2.69 forms a thyristor. The operation can be understood as an interaction of a vertical p-n-p transistor (p-drain, n-well, p-substrate) and a lateral n-p-n transistor (n-well, p-substrate, n-drain).

The current gains are A_{Nnpn} and A_{Npnp}, respectively. If

$$A_{Nnpn} + A_{Npnp} = 1 \qquad (2.216)$$

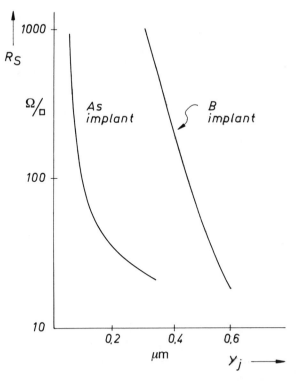

Fig. 2.67. Sheet resistance of drain/source regions.

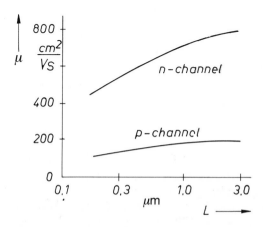

Fig. 2.68. Electron and hole mobility in short-channel transistors.

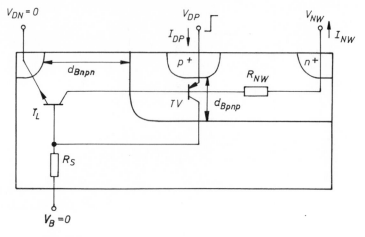

Fig. 2.69. p-n-p-n thyristor structure in CMOS circuits.

or

$$B_{\text{Nnpn}} B_{\text{Npnp}} = 1 \tag{2.217}$$

the thyristor switches in the on state. In this conductive state the whole structure overflows with charge carriers. We will demonstrate this latch-up with results of computer simulations in Figs 2.71–2.74.

In Fig. 2.70 the impurity profile of the n^+-p-n-p^+ structure is plotted. A sudden voltage peak, $V_{\text{DP}} > V_{\text{NW}}$ (see Fig. 2.69), drives the emitter–base junction of the vertical p^+-n-p transistor in the forward direction. This causes a hole injection into the n-well (Fig. 2.71). The voltage drop on the substrate resistor, R_{S}, due to the injection current switches the n^+-p-n lateral transistor into the conducting state leading to an electron injection (Fig. 2.72). Now the current gains A_{Nnpn} and A_{Npnp} are increased until the latch-up condition of Equations (2.216) and (2.217) are fulfilled. The potential and current density distribution under latch-up conditions are shown in Figs. 2.73 and 2.74.

If we supply very short voltage pulses, V_{DP}, with a pulse length T_p which is shorter than the total transit time τ_{T} in the n-p-n and p-n-p transistor

$$T_p < \tau_{\text{T}} = \frac{d_{\text{Bpnp}}^2}{2\mu_p V_{\text{T}}} + \frac{d_{\text{Bnpn}}^2}{2\mu_n V_{\text{T}}}, \tag{2.218}$$

no latch-up occurs. Special epitaxy layers, trench isolation, or Schottky diodes are employed to prevent latch-up (Menozzi *et al.* 1988).

All the effects we have considered in this section are constraints against further device scaling (El Mansy 1982; Lee *et al.* 1988), but it is interesting to note that at low temperatures (e.g. 77 K) nearly all the parameters of MOS transistors are improved (Selberherr 1989).

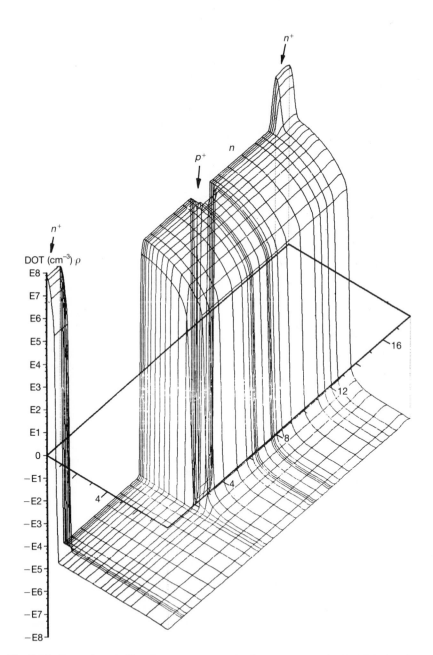

Fig. 2.70. Impurity profile of an n$^+$-p-n-p$^+$ thyristor structure in CMOS circuits.

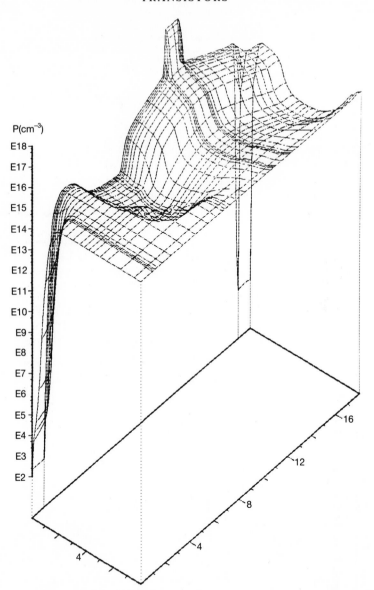

Fig. 2.71. Hole distribution in an n^+-p-n-p^+ thyristor structure of Fig. 2.70 ($V_{DS} = 5.7 \, V$, $V_{GS} = V_{BS} = 0$).

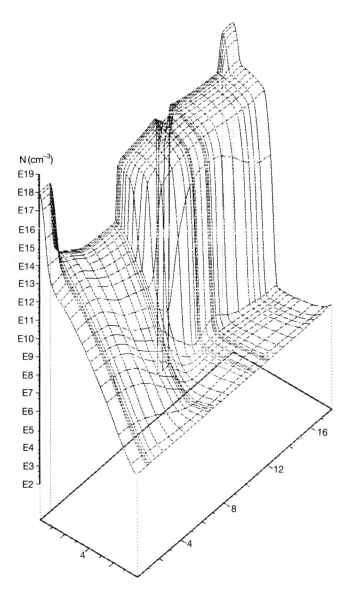

Fig. 2.72. Electron distribution in a n+-p-n-n+ thyristor structure of Fig. 2.70 at latch-up ($V_{DS} = 5.7$ V, $V_{GS} = V_{BS} = 0$).

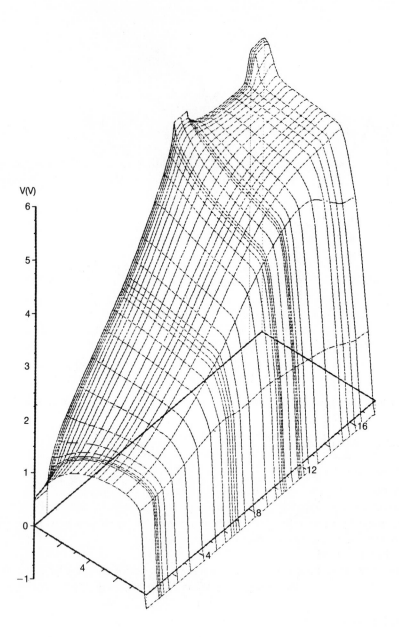

Fig. 2.73. Potential distribution in an n⁺-p-n-p⁺ thyristor structure of Fig. 2.70 at latch-up.

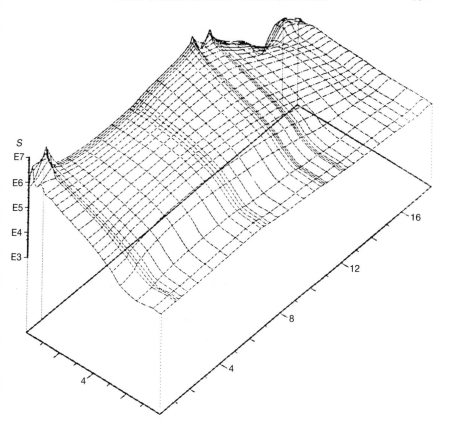

Fig. 2.74. Current distribution in an n+-p-n-p+ thyristor structure of Fig. 2.70.

2.2.5 Network models

Large-signal model

A non-linear network model for computer simulation of MOS transistor circuits is shown in Fig. 2.75. The main model element is the current source I_K given by Equations (2.168), (2.169), and (2.173). It can be summarized by the following model equations.

For $V_{GS} < V_t$

$$I_K = I_o \exp \frac{V_{GS} - V_t}{NV_T} \left(1 - \exp \frac{-V_{DS}}{V_T}\right). \qquad (2.219)$$

For $V_t < V_{GS}$, $V_{DS} \geq K5 (V_{GS} - V_t)$

$$I_K = K1(V_{GS} - V_t)\{(V_{GS} - V_t) + K4V_{DS}\}. \qquad (2.220)$$

For $V_{DS} \leq K5 (V_{GS} - V_t)$

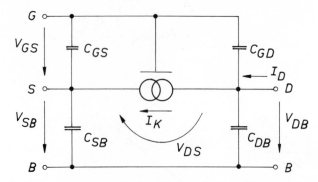

Fig. 2.75. Non-linear network model of the MOS transistor.

$$I_K = K1\{(V_{GS} - V_t)(2 + K4) - V_{DS}\} V_{DS} \qquad (2.221)$$

where

$$K5 \approx 0.8\text{--}1$$

$$V_t = V_{to} + K2\sqrt{2V_F + V_{SB}} \qquad (2.222)$$

$$V_{to} = 2V_F + V_{FB} \qquad (2.223)$$

$$K2 = \frac{d_i}{\epsilon_i}\sqrt{2\, eN_A\epsilon_H} \qquad (2.224)$$

$$K1 = K0 = \tfrac{1}{2}\,\mu_{no}\,\frac{\epsilon_i}{d_i}\,\frac{b}{L}\,. \qquad (2.225)$$

Taking into account the field dependence of the mobility (see Section 2.2.4, Equation (2.209)) and the serial resistors, we have to amend $K1$ to

$$K1 = \frac{K0}{1 + K3(V_{GS} - V_t)}\,. \qquad (2.226)$$

The model parameters for static d.c. operation are $K0$, $K2$, $K3$, $K4$, $K5$, and V_{to}.

For the simulation of the dynamic behaviour the model must be supplemented by capacitors as shown in Fig. 2.75. The physical origin of these capacitors can be seen in Fig. 2.76. C_{DB} and C_{SB} are space-charge layer capacitors of the source–substrate and drain–substrate p-n junctions. Their voltage dependence can be written as

$$C_{DB} = C''_{sdbo} A_D \left(1 + \frac{V_{DB}}{V_{bi}}\right)^q \qquad (2.227)$$

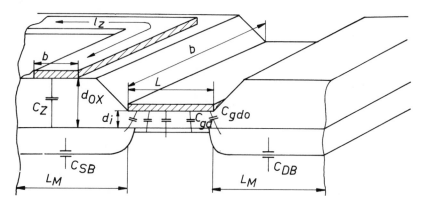

Fig. 2.76. Capacitances at MOS structures.

$$C_{SB} = C''_{ssbo} A_S \left(1 + \frac{V_{SB}}{V_{bi}}\right)^q \qquad (2.228)$$

where $q = 0.3\text{-}0.5$, $V_{bi} \approx 1\,\text{V}$, $A_D = L_M b_M$, $A_S = L_M b_M$, and $C''_{SO} = \sqrt{(e\epsilon_H N_A/2V_{bi})}$.

The capacitances C_{GS} and C_{GD} have two components:

$$C_{GS} = C_{gso} + C_{gs} \qquad (2.229)$$

$$C_{GD} = C_{gdo} + C_{gd} \qquad (2.230)$$

where the capacitors C_{gso} and C_{gdo} are stray capacitors as shown in Fig. 2.76. The inner parts C_{gd} and C_{gs} are voltage dependent and are given here without proof (Möschwitzer and Lunze 1990)

$$C_{gd} = \tfrac{2}{3} C_i \left[1 - \frac{(V_{GS} - V_t)^2}{\{2(V_{GS} - V_t) - V_{DS}\}^2}\right] \qquad (2.231)$$

$$C_{gs} = \tfrac{2}{3} C_i \left[\frac{(V_{GS} - V_t)\{3(V_{GS} - V_t) - 2V_{DS}\}}{\{2(V_{GS} - V_t) - V_{DS}\}^2}\right] . \qquad (2.232)$$

In the active mode of operation, $V_{DS} < V_{GS} - V_t$, we obtain

$$C_{gd} = C_{gs} = \tfrac{1}{2} C_i = \tfrac{1}{2} \frac{\epsilon_i}{d_i} bL . \qquad (2.233)$$

In the pinch-off region, $V_{DS} > V_{GS} - V_t$, we have

$$C_{gd} = 0 \qquad (2.234)$$

$$C_{gs} = \tfrac{2}{3} C_i . \qquad (2.235)$$

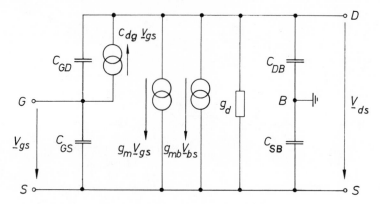

Fig. 2.77. Small-signal equivalent circuit of the MOS transistor.

Small-signal model

The small-signal model Fig. 2.77 can be derived from the large-signal model. The small-signal currents i and voltages v càn be considered as differentials dI and dV of the d.c. currents and voltages (see also Section 2.1.6). We find the following small-signal network parameters.

1. The transconductance is

$$g_m = \frac{\partial I_K}{\partial V_{GS}}\bigg|_{V_{DS}, V_{SB} = \text{const}} \cdot \tag{2.236}$$

For $V_{GS} - V_t \leq V_{DS}$ we have

$$g_m = 2K1(V_{GS} - V_t) + K1K4V_{DS} \approx 2K1(V_{GS} - V_t) \tag{2.237}$$

and for $V_{GS} - V_t \geq V_{DS}$

$$g_m = K1(2 + K4)V_{DS} \approx 2K1V_{DS} \tag{2.238}$$

where $K4 \ll 1$. The decrease of g_m at high gate voltages is due to the decrease of the mobility of charge carriers at the surface (Wong *et al.* 1987) (Fig. 2.78).

2. The back gate transconductance is

$$g_{mb} = \frac{\partial I_K}{\partial V_{BS}}\bigg|_{V_{GS}, V_{DS} = \text{const}} \cdot \tag{2.239}$$

For $V_{GS} - V_t \leq V_{DS}$ we have

$$g_{mb} \approx \frac{K1K2(V_{GS} - V_t)}{\sqrt{2V_F + V_{SB}}} = \lambda g_m \tag{2.240}$$

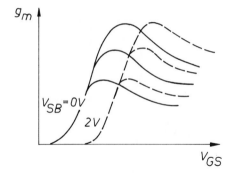

Fig. 2.78. Gate voltage dependence of the small-signal transconductance.

and for $V_{GS} - V_t \geq V_{DS}$

$$g_{mb} \approx \frac{K1K2V_{DS}}{\sqrt{2V_F + V_{SB}}} = \lambda g_m \qquad (2.241)$$

where

$$\lambda = \frac{K2}{2\sqrt{2V_F + V_{SB}}} < 1 .$$

3. The small-signal a.c. drain conductance is

$$g_d = \frac{\partial I_K}{\partial V_{DS}} \bigg|_{V_{DS}, V_{SB} = \text{const}} \qquad (2.242)$$

For $V_{GS} - V_t \leq V_{DS}$ we have

$$g_d = K1K4(V_{GS} - V_t) \qquad (2.243)$$

and for $V_{GS} - V_t \geq V_{DS}$

$$g_d = K1\{(2 + K4)(V_{GS} - V_t) - 2V_{DS}\} \approx 2K1(V_{GS} - V_t - V_{DS}) . \qquad (2.244)$$

The capacitances in Fig. 2.76 are the same as in Fig. 2.77 (Equations (2.227)–(2.235)).

A very detailed analysis gives an additional current $j\omega C_{dg}\mathbf{V}_{gs}$ (in pinch-off $C_{gd} = 0$, but $C_{dg} \neq 0$) in the pinch-off region. It should also be mentioned that we use, in the small-signal model of Fig. 2.77, the complex phasors $\mathbf{I}$ and $\mathbf{V}$ for the a.c. currents and voltages instead of the real time varying small-signal a.c. currents and voltages i and v.

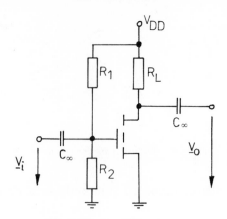

Fig. 2.79. Amplifier stage.

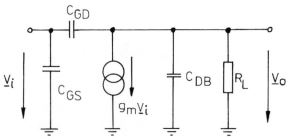

Fig. 2.80. Small-signal equivalent circuit of the amplifier stage of Fig. 2.79.

2.2.6 High-frequency behaviour

For the investigation of the high-frequency behaviour we employ the simple MOS amplifier stage illustrated in Fig. 2.79, where R_1 and R_2 determine the operating point but do not influence the high-frequency behaviour. Because the substrate is grounded the current source, $g_{mb}\mathbf{V}_{bs}$, is zero. We assume that the transistor is always in pinch-off so that we can neglect g_d and C_{gd} and obtain the simplified equivalent circuit shown in Fig. 2.80.

The voltage gain is

$$v = \frac{\mathbf{V}_o}{\mathbf{V}_i} = -\frac{g_m R_L\left(1 + j\omega\dfrac{C_{GD}}{g_m}\right)}{(1 + j\omega C_{DB}R_L) + j\omega C_{GD}R_L}. \tag{2.245}$$

For frequencies that are not too high we have $\omega C_{GD} \ll g_m$ and we find approximately

$$v = \frac{\mathbf{V}_o}{\mathbf{V}_i} = -\frac{g_m R_L}{1 + j\omega R_L(C_{DB} + C_{GD})} \tag{2.246}$$

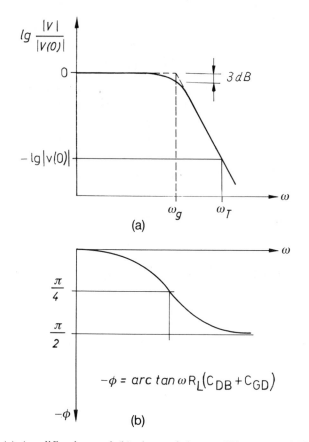

Fig. 2.81. (a) Amplification and (b) phase of the amplifier stage of Fig. 2.79 as a function of frequency.

$$\log \frac{|v|}{|v(0)|} = -\tfrac{1}{2} \log \left(1 + \omega^2 R_{\mathrm{L}}^2 (C_{\mathrm{DB}} + C_{\mathrm{GD}})^2\right). \qquad (2.247)$$

This is depicted in Fig. 2.81(a) (known as a Bode diagram) where

$$|v(0)| = g_{\mathrm{m}} R_{\mathrm{L}} \qquad (2.248)$$

is the low-frequency voltage gain. The phase response is

$$-\varphi = \arctan \omega R_{\mathrm{L}} (C_{\mathrm{DB}} + C_{\mathrm{GD}}) \qquad (2.249)$$

and is sketched in Fig. 2.81(b).

The cut-off frequency, f_{g}, is defined by

$$|v(f_{\mathrm{g}})| = \frac{v(0)}{\sqrt{2}} \qquad (2.250)$$

and is

$$f_g = \frac{\omega_g}{2\pi} = \frac{1}{2\pi R_L (C_{DB} + C_{GD})}. \qquad (2.251)$$

The frequency ω_T at which $|v(\omega_T)|$ is unity is called the transit frequency or gain–bandwidth product. With Equation (2.246) we have

$$\omega_T = 2\pi f_T = \frac{\sqrt{g_m^2 R_L^2 - 1}}{R_L (C_{DB} + C_{GD})} \approx \frac{g_m}{C_{DB} + C_{GD}}. \qquad (2.252)$$

The dynamic input capacitance (Miller capacitance) is

$$C_i = C_{GS} + C_{GD}(1 - v) \qquad (2.253a)$$

or, using $v \approx v(0) = -g_m R_L$,

$$C_i = C_{GS} + C_{GD}(1 + g_m R_L). \qquad (2.253b)$$

The differential amplification of the CMOS amplifier stage in Fig. 2.82 is

$$v_D = \frac{V_o}{V_{i2} - V_{i1}} = \frac{g_{m1}}{g_{d1} + g_{d2}} \qquad (2.254)$$

where g_{m1} and g_{d1} are the small-signal parameters (transconductance and

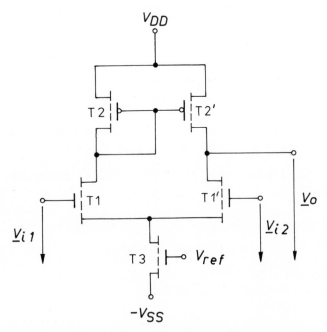

Fig. 2.82. CMOS differential amplifier stage.

a.c. small-signal conductance, see Fig. 2.77) of the n-channel transistor T1, and g_{d2} is the a.c. small-signal conductance of the p-channel transistor T2. The body-effect is eliminated in this CMOS differential amplifier because the p-substrate and the n-well are grounded.

2.2.7 Switching behaviour

The basic circuit of the MOS transistor switch is shown in Fig. 2.83. The transistor is switched on with a high input voltage, $V_{EH} > V_t$, and is switched off with a low input voltage, $V_{EL} < V_t$. This is also depicted in Fig. 2.84. In the on state we have a current, I_{D1}, and a low output voltage, V_{AL}. In the off state the current, I_D, is zero and we have a high output voltage, $V_{AH} \approx V_{DD}$.

The load resistor, R_L, and the transistor aspect ratio b/L must be

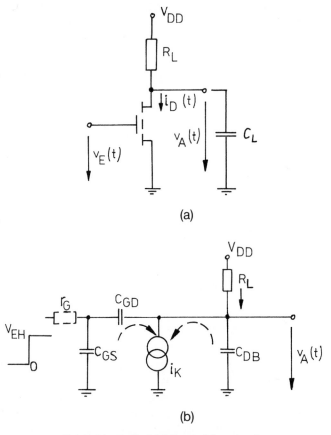

(a)

(b)

Fig. 2.83. Basic MOS switching circuit.

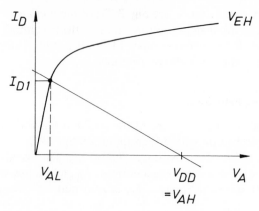

Fig. 2.84. Drain current characteristic showing the on and off state of the MOS transistor switch.

designed so that the low-level of the output voltage V_{AL} is much smaller than the threshold voltage V_t. Using the simple current equations (2.166) we have, for high input voltages V_{EH} and low output voltages in the active mode,

$$I_D = K1(2(V_{EH} - V_t)V_{AL} - V_{AL}^2) \approx 2K1(V_{EH} - V_t)V_{AL} \quad (2.255)$$

where

$$K1 = \tfrac{1}{2}\,\mu_n\,\frac{\epsilon_i}{d_i}\,\frac{b}{L} = K'\,\frac{b}{L}.$$

Further we find

$$I_D = \frac{V_{DD} - V_{AL}}{R_L}. \quad (2.256)$$

From Equations (2.255) and (2.256) we derive

$$V_{AL} = \frac{V_{DD}}{1 + 2R_L K'\dfrac{b}{L}(V_{EH} - V_t)} \approx \frac{L}{bR_L}\,\frac{V_{DD}}{2K'(V_{EH} - V_t)}. \quad (2.257)$$

For $V_{AL} \ll V_t$ it must be the case that

$$\frac{V_{DD}}{2K'(V_{EH} - V_t)} \ll \frac{R_L b}{L}. \quad (2.258)$$

The dynamic behaviour will be investigated in Fig. 2.85. At $t = 0$ the output voltage, V_{AH}, may be high and the transistor is switched on by a high input voltage V_{EH}. Now the transistor current i_K discharges the load capacitor at the output $C_L = C_{DB} + C_{GD}$ (Fig. 2.83(b)). Therefore we have

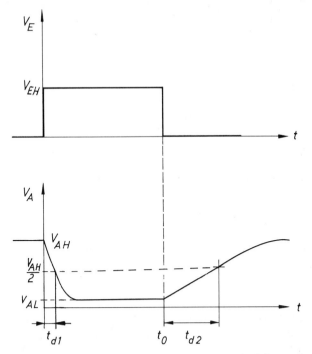

Fig. 2.85. Response of the input and output voltage at a MOS transistor switch.

$$i_K = -C_L \frac{dv_A}{dt} \qquad (2.259)$$

where we have neglected the load current through R_L. We can do this without significant loss of accuracy because the transistor is much more conductive in the on state than the load resistor R_L.

Initially $v_A > V_{EH} - V_t$ and the transistor is in the pinch-off region. In this case we have

$$K1(V_{EH} - V_t)^2 = -C_L \frac{dv_A}{dt}. \qquad (2.260)$$

At $t = t_1$ the transistor leaves the pinch-off region ($v_A(t_1) = V_{EH} - V_t$) and enters the active region where $v_A < V_{EH} - V_t$. Integrating Equation (2.260) we find, with $v_A(t_1) = V_{EH} - V_t$ and $V_{EH} = V_{DD}$,

$$t_1 = \frac{C_L V_t}{K1(V_{EH} - V_t)^2}. \qquad (2.261)$$

For $t > t_1$ we have

$$K1\{2(V_{EH} - V_t)v_A - v_A^2\} = -C_L \frac{dv_A}{dt}. \qquad (2.262)$$

The solution of this differential equation with the initial condition $v_A(t_1) = V_{EH} - V_t$ gives

$$v_A(t) = \frac{2 \exp - t'/\tau}{1 + \exp - t'/\tau} (V_{EH} - V_t) \qquad (2.263)$$

where $t' = t - t_1$ and

$$\tau = \frac{C_L}{2K1(V_{EH} - V_t)}. \qquad (2.264)$$

The off–on delay time, t_{d1}, is defined by $v_A(t_{d1}) = V_{AH}/2$ (see Fig. 2.85) and we derive, from Equations (2.263) and (2.261) with $V_{EH} = V_{AH} = V_{DD}$,

$$t_{d1} = \frac{C_L}{2K1(V_{DD} - V_t)} \left(\frac{2V_t}{V_{DD} - V_t} + \ln \frac{3V_{DD} - 4V_t}{V_{DD}} \right). \qquad (2.265)$$

For common values of $V_t \approx 1$ V and $V_{DD} = 5$ V we obtain approximately

$$t_{d1} = \frac{C_L}{2K1(V_{DD} - V_t)}. \qquad (2.266)$$

At $t = t_o$ the transistor is switched off by a low input voltage $V_{EL} < V_t$. Now the capacitor C_L at the output node is charged by a load current via R_L and we have

$$\frac{V_{DD} - v_A}{R_L} = C_L \frac{dv_A}{dt}. \qquad (2.267)$$

With the initial condition

$$v_A(t_o) = V_{AL} \qquad (2.268)$$

we derive the time response of the output voltage

$$v_A(t) = (V_{DD} - V_{AL}) \left(1 - \exp - \frac{t - t_o}{C_L R_L} \right) + V_{AL}. \qquad (2.269)$$

The on–off delay time is defined by $v_A(t - t_o = t_{d2}) = V_{DD}/2$ (see Fig. 2.85) and we derive, from Equation (2.269)

$$t_{d2} = C_L R_L \ln \frac{2(V_{DD} - V_{AL})}{V_{DD}}. \qquad (2.270)$$

Because of Equation (2.258) we have $t_{d2} \gg t_{d1}$.

In integrated circuits R_L is replaced by another MOS transistor (n-channel depletion-type or p-channel enhancement-type). This will be considered in Chapter 3.

2.2.8 Physical implementation

In Fig. 2.86 we show a cross-section and a layout of a typical MOS transistor structure. The side-wall isolation is achieved by a thick field oxide (LOCOS), and parasitic channels are prevented by p-channel stopper implants. The most important parameters for the designer are the channel length L and the channel width b. L and b determine the transistor constant $K1$ in Equations (2.166)–(2.174) and therefore all other transistor parameters. L and b are calculated by the circuit designer (see Chapter 3). However the mask layout drawing A and B (see Fig. 2.86) must be somewhat larger because of the shortening, ΔL, of all dimensions owing to technological constraints such as underetching and side-wall diffusion. Therefore we have

$$\frac{b}{L} = \frac{A - \Delta L}{B - \Delta L}. \tag{2.271}$$

For the layout a number of design rules must be observed as we have already mentioned in Section 2.1.10. Some of the most important design rules for a MOS process are listed below (see Fig. 2.86):

(1) B minimum gate length;

(2) D minimum width of interconnection lines;

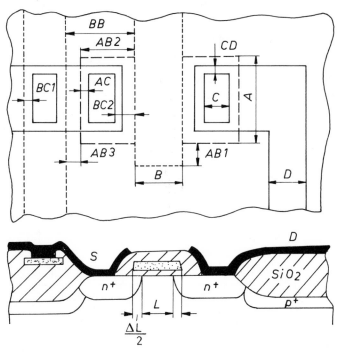

Fig. 2.86. Layout and cross-section of a MOS transistor showing the design rules.

(3) *AB*1 minimum overlap of the polysilicon gate on the active transistor area;

(4) *C* minimum contact window;

(5) *BC*2 minimum distance between contact window and gate.

2.3 GaAs FIELD-EFFECT TRANSISTORS

GaAs, as a basic material for electronic devices, has some important advantages. Among these are (Abe 1982; Shur 1978):

(1) enlarged temperature range and better radiation hardness due to a larger forbidden energy gap and a lower intrinsic density (this is also the reason why a better isolation of individual devices in GaAs integrated circuits is possible);

(2) higher operating speed because of high electron mobility and drift velocities.

Two disadvantages are:

(1) higher process complexity and costs;

(2) sensitivity to mechanical stress.

2.3.1 Fundamentals of junction field-effect transistors

The classical junction field-effect transistor (JFET) is shown in Fig. 2.87. This device has a heavily doped p-gate (G) on an n-epitaxy layer. The n-epitaxy layer acts as an n-channel between the heavily doped n$^+$-drain (D)

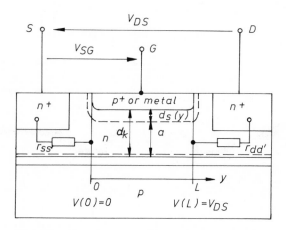

Fig. 2.87. Junction field-effect transistor (JFET).

and source (S). If we supply a negative voltage V_{SG} the gate–channel p-n junction will be reverse biased. The space-charge layer width of this one-sided p-n junction is given by (see Section 1.4.2, Equation (1.54c))

$$d_s = \sqrt{\frac{2\epsilon_H}{eN_D}(V_R + V_{bi})} \tag{2.272}$$

where V_{bi} is the built-in voltage, and V_R is the reverse voltage on the p-n junction along the channel. $V_R = V_{SG} + V(y)$ *and* N_D is the channel doping concentration.

The electronic channel depth is therefore

$$a = d_K - \sqrt{\frac{2\epsilon_H}{eN_D}(V_R + V_{bi})} \tag{2.273}$$

which is voltage dependent. By controlling the electronic channel depth it is possible to control the transistor current I_D. For this current we have

$$I_D = \kappa AE \tag{2.274}$$

where

$$\kappa = e\mu_n N_D \tag{2.275}$$

is the channel conductivity and

$$A = a \times b \tag{2.276}$$

is the channel cross-section (see Fig. (2.87)). The channel field is given by

$$E = \frac{dV}{dy}. \tag{2.277}$$

The voltage drop along the channel has the boundary values $V(0) = 0$ and $V(L) = V_{DS}$ (see Fig. 2.87).

If we insert a of Equation (2.273) and E of Equation (2.277) into Equation (2.274) we obtain

$$I_D = \kappa b d_k \left(1 - \sqrt{\frac{V_{bi} + V(y) + V_{SG}}{V_{bi} + V_p}}\right)\frac{dV}{dy}. \tag{2.278}$$

In this equation we have introduced a pinch-off voltage

$$V_p = \frac{ed_k^2 N_D}{2\epsilon_H} - V_{bi}. \tag{2.279}$$

This is the reverse voltage which must be applied at the gate–channel p-n junction to make the electronic channel depth $a = 0$.

Integrating Equation (2.278) in the interval $y = 0$ ($V(0) = 0$) to $y = L$ ($V(L) = V_{DS}$) yields the I/V characteristic of the JFET

$$I_D = \kappa \frac{d_k b}{L} \left(V_{DS} - \tfrac{2}{3}(V_p + V_{bi}) \left\{ \left(\frac{V_{SG} + V_{DS} + V_{bi}}{V_p + V_{bi}} \right)^{3/2} - \left(\frac{V_{SG} + V_{bi}}{V_p + V_{bi}} \right)^{3/2} \right\} \right)$$

$$(2.280)$$

For $V_{SG} \geq V_p$ the channel is completely pinched off and the current I_D is zero. For $V_{DS} + V_{SG} > V_p$ the channel is partly pinched off. Therefore the I/V characteristic in Equation (2.280) is valid only for $V_{DS} \leq V_p - V_{SG}$.

At $V_{DS} = V_p - V_{SG}$ the pinch-off point starts at the drain and moves for $V_{DS} > V_p - V_{SG}$ toward the source. We find the I/V characteristic in the pinch-off region ($V_{DS} > V_p - V_{SG}$) if we set, in Equation (2.280), V_{DSS} instead of V_{DS}. V_{DSS} is a 'saturation' voltage and can be calculated as follows.

The electric field in the channel can be derived from Equation (2.278) and with I_D from Equation (2.280). If we set $E = E_g$ at the pinch-off point, $V_{DS} = V_{DSS}$, for which the velocity of electrons in the channel reaches its maximum v_g (e.g. $E_g \approx 10^4 \text{ V cm}^{-1}$ for silicon, see Fig. 1.12) we obtain

$$E = \frac{V_{DSS}\left[1 - \dfrac{2(V_{bi} + V_p)}{3V_{DSS}} \left\{ \left(\dfrac{V_{SG} + V_{DSS} + V_{bi}}{V_{bi} + V_p} \right)^{3/2} - \left(\dfrac{V_{SG} + V_{bi}}{V_{bi} + V_p} \right)^{3/2} \right\} \right]}{L\left\{ 1 - \left(\dfrac{V_{SG} + V_{DSS} + V_{bi}}{V_{bi} + V_p} \right)^{1/2} \right\}} .$$

$$(2.281)$$

From Equation (2.281) we can derive a simple approximation for the saturation voltage V_{DSS}:

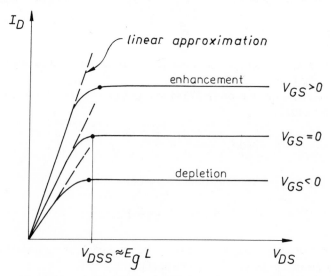

Fig. 2.88. Simplified current–voltage characteristics of a junction field-effect transistor.

$$V_{\mathrm{DSS}} \approx \frac{(V_{\mathrm{p}} - V_{\mathrm{SG}})LE_{\mathrm{g}}}{LE_{\mathrm{g}} + (V_{\mathrm{p}} - V_{\mathrm{SG}})}. \tag{2.282}$$

For long-channel transistors ($L \gg 10\,\mu$m) we have $V_{\mathrm{DSS}} = V_{\mathrm{p}} - V_{\mathrm{SG}}$ and for short-channel transistors ($L \ll 2\,\mu$m) we have $V_{\mathrm{DSS}} = E_{\mathrm{g}}L$.

Fig. 2.88 shows the output characteristics. For $V_{\mathrm{DS}} \ll V_{\mathrm{SG}} + V_{\mathrm{bi}}$ we have the linear approximation

$$I_{\mathrm{D}} = \kappa \frac{d_{\mathrm{k}}b}{L} \left\{ 1 - \left(\frac{V_{\mathrm{SG}} + V_{\mathrm{bi}}}{V_{\mathrm{bi}} + V_{\mathrm{p}}} \right)^{\frac{1}{2}} \right\} V_{\mathrm{DS}}. \tag{2.283}$$

2.3.2 MESFET

The MESFET (metal–semiconductor field-effect transistor) is a junction field-effect transistor with a Schottky gate (Das 1987; Taylor 1987). In Fig. 2.89 a cross-section and a layout is shown. The transistor is made on semi-insulating (SI) GaAs substrates. The n-channel is separated by a buffer

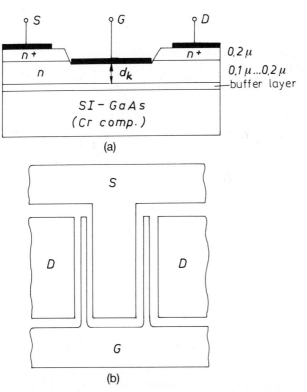

Fig. 2.89. Cross-section and layout of a MESFET.

layer from the SI substrate. The transistor current is controlled by the space-charge layer width of the metal–semiconductor Schottky diode and is given by Equations (2.280) and (2.283) and is plotted in Fig. 2.88.

The pinch-off voltage of Equation (2.279)

$$V_p = \frac{ed_k^2 N_D}{2\epsilon_H} - V_{bi} \tag{2.284}$$

is determined by the built-in voltage, V_{bi}, of the Schottky diode (see Equation (1.121))

$$V_{bi} = V_B - V_T \ln \frac{N_C}{N_D} \tag{2.285}$$

where V_B is the Schottky barrier height and N_C is the effective density of states in the conduction band. Normally V_p is positive and the MESFET is a depletion-type transistor.

For small channel thickness

$$d_k < \sqrt{\frac{2 \epsilon_H \left(V_B - V_T \ln \frac{N_C}{N_D} \right)}{e N_D}} \tag{2.286}$$

the pinch-off voltage V_p is negative ($V_p = -V_{pEM}$) and we get an enhancement-type MESFET with buried channel. This means the channel is fully depleted at $V_{GS} = 0$ and, for $V_{GS} > V_{pEM}$, a channel will be enhanced. In Figs. 2.90 and 2.91 measured transfer characteristics for depletion- and enhancement-type MESFETs are plotted.

The small-signal high-frequency model is shown in Fig. 2.92 where g_m is

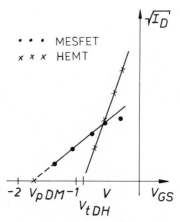

Fig. 2.90. Transfer characteristics of depletion-type MESFET and depletion-type HEMT.

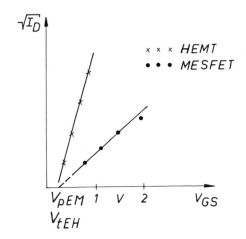

Fig. 2.91. Transfer characteristics of enhancement-type MESFET and enhancement-type HEMT.

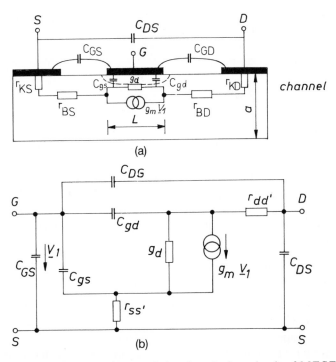

Fig. 2.92. (a) Model and (b) small-signal equivalent circuit of MESFET.

the transconductance. In the active mode of operation $V_{DS} < V_{DSS}$ we have with Equation (2.283)

$$g_m = \frac{-\partial I_D}{\partial V_{SG}} \bigg|_{V_{DS}} = \kappa \frac{d_k b}{L} \frac{V_{DS}}{2\{(V_p + V_{bi})(V_{bi} + V_{SG})\}^{1/2}}. \quad (2.287)$$

In the pinch-off region ($V_{DS} > V_{DSS}$) we have, with Equation (2.280) and $V_{DS} = V_{DSS}$,

$$g_m = \frac{-\partial I_D}{\partial V_{SG}} \bigg|_{V_{DSS}} = \kappa \frac{d_k b}{L} \left\{ \left(\frac{V_{SG} + V_{DSS} + V_{bi}}{V_p + V_{bi}} \right)^{1/2} - \left(\frac{V_{SG} + V_{bi}}{V_p + V_{bi}} \right)^{1/2} \right\}$$

$$(2.288)$$

or, with $V_{DSS} = V_p - V_{SG}$ for long-channel transistors,

$$g_m = \kappa \frac{d_k b}{L} \left\{ 1 - \left(\frac{V_{SG} + V_{bi}}{V_p + V_{bi}} \right)^{1/2} \right\}. \quad (2.289)$$

The small-signal a.c. conductance is, in the active mode of operation,

$$g_d = \frac{\partial I_D}{\partial V_{DS}} \bigg|_{V_{SG}} = \kappa \frac{d_k b}{L} \left\{ 1 - \left(\frac{V_{SG} + V_{bi}}{V_p + V_{bi}} \right)^{1/2} \right\}. \quad (2.290)$$

The drain and source resistors $r_{dd'}$ and $r_{ss'}$ are determined by contact (r_K) and spreading resistors (r_B) as shown in Fig. 2.92(a).

The capacitors C_{gs} and C_{gd} are space-charge layer capacitors of the Schottky diode and can be calculated for $V_{DS} < V_{SG}$ by (Shur 1978)

$$C_{gs} = C_{gd} = \frac{\epsilon_H bL}{2d_k} \sqrt{\frac{V_p + V_{bi}}{V_{SG} + V_{bi}}}. \quad (2.291)$$

In the pinch-off region C_{gd} is very small ($C_{gd} \approx 0$) and C_{GS}, C_{GD}, and C_{DS} are stray capacitors between the drain, gate, and source electrodes.

MESFETs are applied as microwave amplifiers up to 100 GHz and as ultra high-speed switches in modern digital systems such as supercomputers (Das 1987; Long 1989; Taylor 1987). For typical gate lengths of $L = 0.5\,\mu m$, signal delay times of the order of 10 ps and transconductances of 400 mS/mm are possible. MESFET timer and counter circuits can run faster than 10 GHz.

2.3.3 HEMT

HEMT means high electron mobility transistor (Abe *et al.* 1989; Chang and Fetterman 1987; Delagebeaubeuf and Lenk 1982). Sometimes it is called modulation-doped field-effect transistor (Chung *et al.* 1989; Lee *et al.* 1983). Its operation is based on a two-dimensional electron gas with extremely high mobility ($\mu_n > 10,000\,cm^2/Vs$, see Fig. 2.96). This electron gas is captured

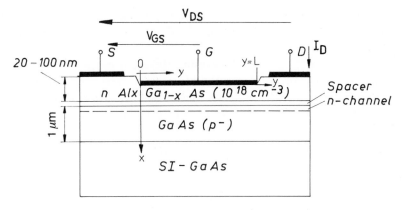

Fig. 2.93. HEMT.

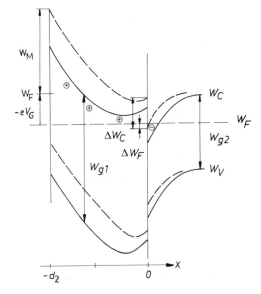

Fig. 2.94. Energy-band diagram of depletion-type HEMT (solid $V_G = 0$; dashed $V_G > 0$)

at the interface of a heterojunction, such as AlGaAs–GaAs as shown in Fig. 2.93. The energy-band diagram of the AlGaAs–GaAs heterojunction is sketched in Fig. 2.94. There is a discontinuity in the conduction band and valence band due to different values of the electron affinity W_{EA} in AlGaAs and GaAs, respectively. Because of the smaller electron affinity in AlGaAs we have an energy minimum in the conduction band of GaAs. In this minimum an electron gas is captured. This minimizes the interaction of the electrons with the impurities in the doped AlGaAs and therefore keeps the mobility in the undoped GaAs high. The scattering can be further minimized by employing a spacer ($d < 20\,\text{nm}$) in between the n-doped AlGaAs and the undoped GaAs.

We derive the I/V characteristic using the same methods as we have already done in Section 2.2.2. for the MOS field-effect transistor (Chang and Fetterman 1987; Delagebeaubeuf and Lenk 1982; Lee *et al.* 1983). The transistor current is a drift current of the high mobility electron gas in the y direction

$$I_D = \mu_n(-Q_n'')\,b\,\frac{dV}{dy} \tag{2.292}$$

where μ_n is the high electron mobility in the undoped GaAs ($\mu_n = \mu_{n1}$). Q_n'' is the electron charge per unit area in the two-dimensional electron gas and is given by Gauss's law (see Fig. 2.94)

$$-Q_n'' = \epsilon_{H2} E(x = -0) \tag{2.293}$$

where ϵ_{H2} is the permittivity in the AlGaAs layer.

The electric field at $x = -0$, $E(-0)$, can be evaluated by integration of Poisson's equation in $-d_2 \le x \le 0$ (AlGaAs) twice so that

$$\frac{d^2\varphi}{dx^2} = -\frac{e}{\epsilon_{H2}} N_{D2} \tag{2.294}$$

where N_{D2} is the donor concentration in AlGaAs.

With the boundary conditions (see Fig. 2.94(a)) $\varphi(0) = 0$, $\varphi(-d_2) = -\{V_M - V_G - (\Delta V_C - \Delta V_F)\} \approx -(V_M - V_G - \Delta V_C)$, and $V_G = V_{GS} - V(y)$, the integration of Poisson's equation yields

$$-\{V_M - (V_{GS} - V(y)) - \Delta V_C\} = -\frac{e}{\epsilon_{H2}} N_{D2}\frac{d_2^2}{2} + E(-0)d_2 \tag{2.295}$$

where V_M is the metal work function (in volts), ΔV_C is the discontinuity in the conduction band (see Fig. 2.94), and $V(y)$ is the voltage drop along the channel (with respect to the source where we set $V(0) = 0$).

Solving Equation (2.295) for $E(-0)$ gives the channel charge per unit area

$$-Q_n'' = \epsilon_{H2} E(-0) = \frac{\epsilon_{H2}}{d_2}(V_{GS} - V(y) - V_t) \tag{2.296}$$

where V_t is a threshold voltage (Delagebeaubeuf and Lenk 1982)

$$V_t = V_M - \Delta V_C - \frac{e}{\epsilon_{H2}} N_{D2}\frac{d_2^2}{2}. \tag{2.297}$$

Integration of Equation (2.296) gives the I/V characteristic of the HEMT in the active mode of operation ($V_{DS} \leq V_{GS} - V_t$)

$$I_D = \mu_{n1} \frac{\epsilon_{H2}}{d_2} \frac{b}{L} \left\{ (V_{GS} - V_t) V_{DS} - \frac{V_{DS}^2}{2} \right\}$$ (2.298)

and, in the pinch-off region ($V_{DS} > V_{GS} - V_t$),

$$I_D = \mu_{n1} \frac{\epsilon_{H2}}{d_2} \frac{b}{L} \frac{(V_{GS} - V_t)^2}{2}.$$ (2.299)

For

$$d_2 > \sqrt{\frac{2\epsilon_{H2}}{eN_{D2}}} (V_M - \Delta V_C)$$ (2.300)

the threshold voltage is negative, $V_t = -V_{tDH}$, and we get a depletion-type HEMT.

For

$$d_2 < \sqrt{\frac{2\epsilon_{H2}}{eN_{D2}}} (V_M - \Delta V_C)$$ (2.301)

the threshold voltage is positive, $V_t = V_{tEH} > 0$, and we have an enhancement-type HEMT. The energy-band diagram for this case is sketched in Fig. 2.95.

In Fig. 2.90 and Fig. 2.91 the transfer characteristics of MESFET and HEMT are compared. Because of the higher mobility the HEMT has the larger transconductance. The small-signal high-frequency model is the same

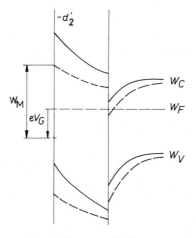

Fig. 2.95. Energy-band diagram of enhancement-type HEMT.

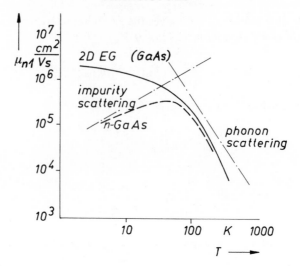

Fig. 2.96. Electron mobility in GaAs and in the two-dimensional electron gas (2D EG) at low temperatures.

as for the MESFET in Fig. 2.92. Using Equations (2.299) and (2.288) we obtain, for the transconductance in the active region ($V_{DS} \leq V_{GS} - V_t$),

$$g_m = \frac{\partial I_D}{\partial V_{GS}}\bigg|_{V_{DS}} = \mu_{n1} \frac{\epsilon_{H2}}{d_2} \frac{b}{L} V_{DS} \qquad (2.302)$$

and, in the pinch-off region ($V_{DS} > V_{GS} - V_t$),

$$g_m = \mu_{n1} \frac{\epsilon_{H2}}{d_2} \frac{b}{L} (V_{GS} - V_t). \qquad (2.303)$$

As an example we have, at 300 K, $g_m' = g_m/b = 200\text{–}500\,\text{mS mm}^{-1}$, and, at 77 K, $g_m' = 400\text{–}1{,}000\,\text{mS mm}^{-1}$; ($\mu_{n1}$ see Fig. 2.96).

2.4 PROBLEMS FOR CHAPTER 2

1. Sketch the energy-band diagram for an n-p-n transistor at thermal equilibrium and for normal active mode of operation.
2. An n-p-n transistor has the doping profile shown in Fig. 2.97 and an emitter area $A_E = 10^{-5}\,\text{cm}^2$. The donor concentration in the emitter is $N_{DE} = 5 \times 10^{19}\,\text{cm}^{-3}$ and the acceptor concentration in the base is $N_{AB} = 10^{18}\,\text{cm}^{-3}$.
   (a) Calculate the emitter saturation current I_{ES}.

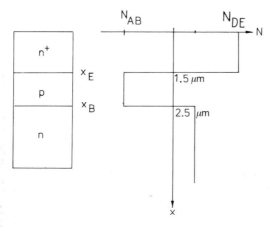

Fig. 2.97. Impurity profile of an n-p-n bipolar transistor.

(b) Calculate the emitter injection efficiency γ_N, the base transport factor, and the common-emitter current gain B_N.

(Use the following parameters: electron mobility in the base, $\mu_{nB} = 2\mu_{pE} = 1{,}000 \; \text{cm}^2/\text{Vs}$; minority carrier lifetime in the base, $\tau_{nB} = 10^{-8} \; \text{s}$; intrinsic density $n_i = 1.5 \times 10^{10} \; \text{cm}^{-3}$; and $V_T = 0.025 \; \text{V}$.)

3. At a bipolar transistor four currents are measured as shown in Fig. 2.98.
   (a) Determine the current parameters I_{ES}, I_{CS}, I_{CES}, and I_{ECS} in the Ebers–Moll model of Fig. 2.8.
   (b) Estimate the normal and inverse current amplification factors A_N, A_1, B_N, and B_1.
4. Derive a formula for ΔV of the transistor circuit in Fig. 2.99 by using the Ebers–Moll model in Fig. 2.8.

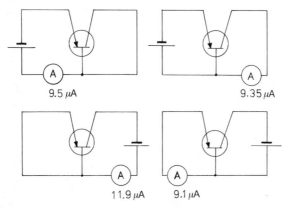

Fig. 2.98. Transistor circuits showing measured currents.

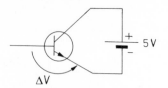

Fig. 2.99. Transistor circuit.

5. Given an n-p-n transistor with a homogeneously doped base of width $d_B = 1\,\mu m$, calculate the minority carrier transit time if the electron mobility is $\mu_n = 800\,cm^2/Vs$. (Use: $kT/e = 0.025\,V$.)
6. The emitter current of an n-p-n transistor is $I_E = 1\,mA$ and the emitter efficiency is $\gamma_N = 0.95$. Calculate the electron charge stored in the homogeneously doped base (of width $d_B = 0.5\,\mu m$) in the normal active mode. (Use: electron mobility, $\mu_n = 1,000\,cm^2/Vs$; and $kT/e = 0.025\,V$.)
7. Find the operating point of the simple bipolar transistor circuit shown in Fig. 2.100(a) and identity it in the collector current characteristic in Fig. 2.100(b).

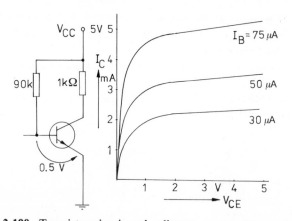

Fig. 2.100. Transistor circuit and collector current characteristics.

8. A basic switching circuit with a bipolar transistor is given in Fig. 2.101. When the transistor is turned on, the emitter–base voltage is constant, $V_{BEX} = V_{FO} = 0.7\,V$, and the collector–emitter saturation voltage is $V_{CES} = 0.1\,V$. The transistor has a common-emitter d.c. current gain of $B_N = 60$ and a charge-storage time constant $\tau_s = 10^{-8}\,s$. Calculate the turn-off delay time t_s (storage time).

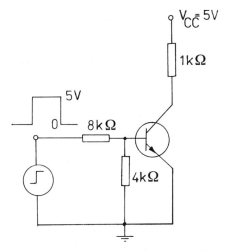

Fig. 2.101. Transistor switching circuit.

9. (a) Derive the frequency dependence of the effective transconductance

$$g_{me} = \frac{I_c}{V_{be}}\bigg|_{V_{ce} = 0}$$

of a bipolar transistor. Its small-signal high-frequency behaviour may be described with the equivalent circuit shown in Fig. 2.102 with $g_m = I_E/V_T$; $r_e = b_n V_T/I_E$; $C_e = C_{se} + C_{de}$; and $V_T = kT/e = 0.025$ V.

(b) Determine the cut-off frequency f_1.

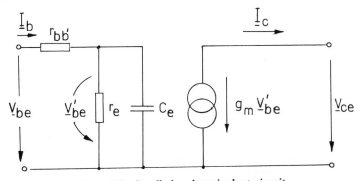

Fig. 2.102. Small-signal equivalent circuit.

10. In Fig. 2.103 an amplifier stage is given with $B_N = 40$.
   (a) Calculate the resistances R_1 and R_L.
   (b) Sketch the low-frequency small-signal equivalent circuit for this stage.

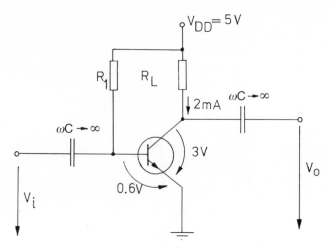

Fig. 2.103. Amplifier stage.

(c) Calculate the a.c. voltage amplification V_o/V_i and the input impedance at low frequencies. The bipolar transistor has the following hybrid parameters: $h_{11e} = 2.1\,\mathrm{k\Omega}$; $h_{12e} = 0$; $h_{21e} = -40$; and $h_{22e} = -120\,\mu\mathrm{S}$.

(d) Derive the a.c. voltage amplification V_o/V_i for $h_{11e} = b_n V_T/I_E$; $h_{12e} = h_{22e} = 0$; and $-h_{21e} = b_n = 40$.

11. Suppose the temperature dependence of the emitter 'saturation' current I_{ES} is given by

$$I_{ES} = 2 \times 10^{-10}\,\mathrm{mA}\,\exp\left(0.15(\Theta/{}^{\circ}\mathrm{C} - 20)\right).$$

and the d.c. current gain is $B_N = 50$, independent of temperature. The supply voltage for a grounded emitter circuit is $V_{CC} = 5\,\mathrm{V}$ and the collector load resistance is $R_L = 200\,\Omega$.

(a) Estimate the operating points $(V_{CE},\ I_C)$ for $\Theta = 20\,^{\circ}\mathrm{C}$ and $\Theta = 50\,^{\circ}\mathrm{C}$, with $V_{BE} = 600\,\mathrm{mV}$.

(b) Derive the temperature feed-through

$$D_T = \left.\frac{\Delta V_{BE}}{\Delta\Theta}\right|_{I_C\,=\,\mathrm{const}}.$$

12. A steady-state current of 1 pA is flowing onto the gate of a MOS transistor. How long would it take until the gate oxide breaks down? The gate area is $A_G = 100\,\mu\mathrm{m}^2$ and the critical field in the gate oxide for breakdown is $E_{crit} = 5 \times 10^6\,\mathrm{V/cm}$. (Use: $\epsilon_{ox} = 3 \times 10^{-13}\,\mathrm{As/Vcm}$.)

13. An n-channel MOS transistor is fabricated in silicon gate technology with a flat-band voltage of $V_{FB} = -1\,\mathrm{V}$, a substrate doping concentration of $N_A = 10^{15}\,\mathrm{cm}^{-3}$, and a gate oxide thickness $d_{ox} = 20\,\mathrm{nm}$. Cal-

culate the threshold voltage for substrate biases of $V_{SB} = 0$ and $V_{SB} = 5$ V and explain the results. (Use: $\epsilon_H = 10^{-12}$ As/Vcm; $\epsilon_{ox} = 3.10^{-13}$ As/Vcm; and $kT/e = 0.025$ V; $n_i = 1.5 \times 10^{10}$ cm^{-3}.)

14. (a) Plot the I/V curve of an enhancement-mode n-channel MOSFET if the source and substrate are grounded and the gate is short-circuited to the drain. (Use $b/L = 1$; $\mu_n = 500$ cm^2/Vs; $\epsilon_{ox}/d_{ox} = 10^{-7}$ As/Vcm^2; and $V_t = 1$ V.)

(b) Repeat part (a) for a depletion-mode n-channel MOSFET of the same size for which the gate is short-circuited to the source ($V_{tD} = -3$ V).

(c) Compare and explain the results.

15. The n-channel enhancement-type MOS transistor T1 in the circuit shown in Fig. 2.104 acts as a transfer gate. It has a threshold voltage including body effect

$$V_{tE}/\text{Volt} = 0.5\{1 + (V_{SB}/\text{Volt})^{0.5}\}.$$

What clock voltage Φ is required at the gate of T1 to get a voltage $V_2 = 4$ V?

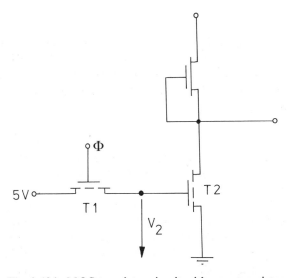

Fig. 2.104. MOS transistor circuit with pass transistor.

16. Assume we have a circuit with two n-channel enhancement-type MOS transistors as shown in Fig. 2.105. Both MOS transistors have the same size and threshold voltage including body effect

$$V_{tE}/\text{Volt} = 0.5\{1 + (V_{SB}/\text{Volt})^{0.5}\}.$$

Find the input voltage V_{i2} on gate 2.

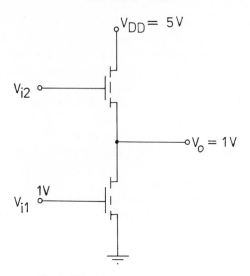

Fig. 2.105. MOS transistor circuit.

17. The resistance $R_D = V_{DS}/I_D$ of two enhancement-type MOS transistors are measured at constant $V_{GS} - V_t$ and low drain voltages $V_{DS} \ll V_{GS} - V_t$. Both transistors have the same parameters; only the projected channel length L_Z is different. R_D is plotted against L_Z and the true channel length is smaller than L_Z:

$$L = L_Z - \Delta L,$$

because of the lateral diffusion of source and drain donors into the channel region during high temperature processing. Use the R_D versus L_Z plot to estimate ΔL.

18. By means of boron implantation the threshold voltage of an n-channel enhancement-type MOS transistor should be increased by $\Delta V_t = 0.5\,\text{V}$. The gate oxide thickness is $d_{ox} = 20\,\text{nm}$ and the permittivity of the oxide is $\epsilon_{ox} = 3 \times 10^{-13}\,\text{Vs/Acm}$. What boron dose (per unit area) is needed to achieve this?

19. In Fig. 2.106 we show a floating-gate transistor (SAMOST). In the programming mode a hot electron current, $I_h = 5\,\text{pA}$, flows from the semiconductor surface through the oxide straight to the floating gate. Estimate the programming time, T_p, which causes a threshold voltage shift of $\Delta V_t = 10\,\text{V}$. The following parameters are given: $d_{ox1} = 30\,\text{nm}$; $\epsilon_{ox} = 3 \times 10^{-13}\,\text{As/Vcm}$; and the area of the floating gate is $A_G = 150\,\mu\text{m}^2$.

20. The circuit diagram of a source follower with two n-channel

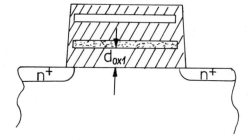

Fig. 2.106. Floating-gate transistor.

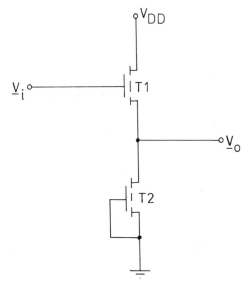

Fig. 2.107. Source follower with MOS transistors.

enhancement-type MOS transistors is given in Fig. 2.107. Use the small-signal equivalent circuit from Fig. 2.77 and derive the formula for the voltage amplification $\mathbf{V}_o/\mathbf{V}_i$ at low frequencies.

21. Calculate and compare the transconductances of a bipolar transistor in active mode and a MOS transistor in pinch-off mode for the same currents $I_C = I_D = 1$ mA. Explain the results. (Use: $I_D = K(V_{GS} - V_t)^2$); $K = 100\ \mu\text{A/V}^2$; and $kT/e = 0.025$ V.)

22. Calculate the pinch-off voltage of a silicon MESFET (see Fig. 2.89) with $d_k = 1\ \mu\text{m}$, $N_D = 10^{15}\ \text{cm}^{-3}$, and $V_B = 0.8$ V. For what channel thickness d_k do we get an enhancement-mode (buried channel) MESFET? (Use: $kT/e = 0.025$ V; $\epsilon_H = 10^{-12}$ As/Vcm; and effective density of states in conduction band, $N_C = 10^{19}\ \text{cm}^{-3}$.)

3 Semiconductor circuits

3.1 BASIC GATES FOR DIGITAL CIRCUITS

The smallest circuit elements of integrated circuits are basic gates which contain only a few devices. Corresponding to the main types of transistors treated in Chapter 2 we have two groups of basic gates in:

(1) silicon and GaAs field-effect technology;

(2) silicon bipolar technology.

3.1.1 Basic gates in silicon MOS technology

NMOS
The NMOS basic gate is the inverter with a load resistor R_L or an n-channel MOS field-effect transistor shown in Fig. 3.1 and 3.2, respectively. The inverter in Fig. 3.1 is used mainly in static memory cells (see Section 4.5.2). In this case the resistance is extremely high ($R_L > 500\,\text{M}\Omega$) and is

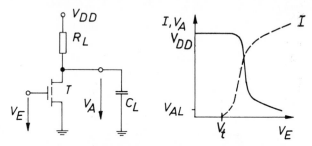

Fig. 3.1. MOS inverter with resistive load.

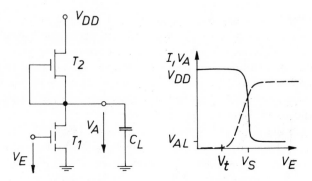

Fig. 3.2. MOS inverter with depletion load.

made of undoped polysilicon layers. The static and dynamic behaviour has already been considered in Section 2.2.7. The transfer characteristic, $V_A = f(V_E)$, and the current, I, (dashed) are plotted in Fig. 3.1(b). If the input voltage, V_E, is smaller than the threshold voltage, V_t, the transistor is switched off and the output voltage has its high-level value $V_{AH} = V_{DD}$.

For $V_A > V_E - V_t$ the transistor is in the pinch-off region and the current is

$$I = K_1 (V_E - V_t)^2. \tag{3.1}$$

Later on, for $V_A < V_E - V_t$, the transistor is in the active region and the current is

$$I = K_1 (2(V_E - V_t)V_A - V_A^2) \tag{3.2}$$

where

$$V_A = V_{DD} - IR_L \tag{3.3}$$

(see Fig. 3.1) and

$$K_1 = \tfrac{1}{2} \mu_n \frac{\epsilon_i b_1}{d_i L_1}$$

(see Section 2.2).

At high input voltage, $V_E = V_{EH}$, the output voltage has its low-level value (see Equation (2.257))

$$V_{AL} = \frac{V_{DD}}{1 + 2R_L K_1 (V_{EH} - V_t)} \approx \frac{V_{DD}}{2R_L K_1 (V_{EH} - V_t)} \tag{3.4}$$

and the current is

$$I_H = \frac{V_{DD} - V_{AL}}{R_L}. \tag{3.5}$$

The dynamic behaviour is determined by the charging and discharging of the load capacitor C_L. Using Equation (2.265) we derive the fall time signal delay (see Fig. 3.3) $V_{EH} = V_{DD}$

$$t_{dl} = \frac{\gamma C_L}{2K_1 (V_{EH} - V_t)} = \tau_1 \frac{C_L}{C_G} \gamma \tag{3.6}$$

where

$$\gamma = \frac{2V_t}{V_{EH} - V_t} + \ln \frac{3V_{EH} - 4V_t}{V_{EH}} \approx 1$$

and

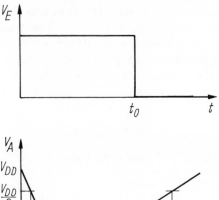

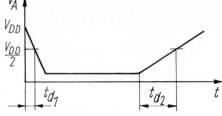

Fig. 3.3. Switching response of input and output voltage of MOS inverters (simplified).

$$\tau_1 = \frac{L^2}{\mu_n (V_{EH} - V_t)} . \qquad (3.7)$$

C_G is the gate capacitance

$$C_G = \frac{\epsilon_i b_1 L_1}{d_i} . \qquad (3.8)$$

For the rise time signal delay we have

$$t_{d2} = C_L R_L \ln \frac{2(V_{DD} - V_{AL})}{V_{DD}} .$$

$$t_{d2} \approx C_L R_L \ln 2. \qquad (3.9)$$

An inverter with an n-channel depletion load transistor has, for $V_E < V_t$, the high-level value of the output voltage, $V_{AH} = V_{DD}$. In this case transistor T_1 is 'off' and T_2 is 'on'. For $V_E > V_t$ T_1 moves into the on state, the output voltage, V_A, decreases and the current, I, increases as shown in Fig. 3.2(b). For $V_A > V_E - V_t$, T_1 is in the pinch-off region and the current is

$$I = K_1 (V_E - V_t)^2. \qquad (3.10)$$

For $V_A < V_E - V_t$, T_1 is in the active region and we have

$$I = K_1 (2(V_E - V_t) V_A - V_A^2). \qquad (3.11)$$

If $V_A < V_{DD} - V_{tD}$, T_2 is in the pinch-off region and it is

$$I_H = K_2 V_{tD}^2 \tag{3.12}$$

where

$$K_2 = \tfrac{1}{2} \mu_n \frac{\epsilon_i b_2}{d_i L_2} \tag{3.13}$$

is the transistor current constant of the depletion transistor T_2. If we supply a high input voltage V_{EH} the output voltage has a low-level value V_{AL}. In that case T_1 is in the active region and T_2 is in the pinch-off region. Therefore Equations (3.11) and (3.12) yield

$$K_2 V_{tD}^2 = K_1 (2(V_{EH} - V_t) V_{AL} - V_{AL}^2) \tag{3.14}$$

and

$$V_{AL} = (V_{EH} - V_t) - \sqrt{(V_{EH} - V_t)^2 - \frac{1}{\alpha} V_{tD}^2} \tag{3.15}$$

where

$$\alpha = \frac{K_1}{K_2} = \frac{(b_1/L_1)}{(b_2/L_2)} \tag{3.16}$$

is the aspect ratio of the transistors. It is necessary for digital circuits that the low-level voltage at the output V_{AL} is much smaller than the threshold voltage V_t to keep the adjacent stage off. Therefore α must be much larger than unity (e.g. $\alpha = 5\text{–}8$ (Carr and Mize 1972; Penney and Lau 1972)). For example if $V_{EH}/V_t = 5$, $\alpha = 5$, and $V_{tD}/V_t = 3$ we have

$$\frac{V_{AL}}{V_t} = 4 - \sqrt{16 - \frac{9}{5}} = 0.23. \tag{3.17}$$

which is low enough to keep the adjacent stage off.

A switching or threshold point, V_S, in the transfer characteristic is defined as shown in Fig. 3.2(b), where a step junction from high-level to low-level output voltage occurs. In this case both transistors T_1 and T_2 are in the pinch-off region and we obtain

$$V_S = V_t + \frac{1}{\sqrt{\alpha}} |V_{tD}|. \tag{3.18}$$

Let us now consider the dynamic behaviour. The time response during the fall time is the same as for the basic gate in Fig. 3.1. If we neglect the weak pull-up action of the small transistor T_2 again and consider only the discharge of the load capacitor C_L via the much larger switching transistor T_1

($\alpha \gg 1$) we have the same results for the fall time signal delay as in Equation (3.6). We repeat it here for convenience:

$$t_{d1} = \tau_1 \frac{C_L}{C_G} \gamma \approx \tau_1 \frac{C_L}{C_G}$$

where

$$\gamma = \frac{2V_t}{V_{EH} - V_t} + \ln \frac{3V_{EH} - 4V_t}{V_{EH}} \approx 1$$

and

$$\tau_1 = \frac{L_1^2}{\mu_n (V_{EH} - V_t)}$$

$$C_G = \frac{\epsilon_i b_1 L_1}{d_i}.$$

The rise time signal delay, t_{d2}, is determined by the charging of the load capacitor C_L via the transistor T_2 (T_1 is now off). If we assume that T_2 is in the pinch-off region during the whole rise time we have

$$K_2 V_{tD}^2 = C_L \frac{dv_A}{dt}. \tag{3.19}$$

The solution of Equation (3.19) with the initial conditions $v_A(t_0) = V_{AL} \approx 0$ and $v_A(t - t_0 = t_{d2}) = V_{DD}/2$ (see Fig. 3.3) yields

$$t_{d2} = \frac{C_L V_{DD}}{2K_2 V_{tD}^2} = 2\alpha \, \tau_2 \frac{C_L}{C_G} \tag{3.20}$$

with

$$\tau_2 = \frac{L_1^2 V_{DD}}{2\mu_n V_{tD}^2} \tag{3.21}$$

and the gate capacitance

$$C_G = \frac{\epsilon_i b_1 L_1}{d_i}.$$

The ratio of the rise and fall time delays is

$$\frac{t_{d2}}{t_{d1}} = 2\alpha \frac{\tau_2}{\tau_1} \approx 2\alpha \gg 1 \tag{3.22}$$

where, because $V_{EH} - V_t \approx 2V_{tD}^2/V_{DD} \approx 4$ volts, we set $\tau_2 \approx \tau_1$.

The dynamic behaviour is determined mainly by the signal delay time t_{d2}

due to the aspect ratio $\alpha \gg 1$ which is needed to keep the low-level voltage V_{AL} much smaller than the threshold voltage V_t. The time response depends on $\dfrac{L^2}{\mu_n}$, α, and C_L/C_G (see Equation (3.20))

$$t_d \sim \frac{L^2}{\mu_n} \frac{C_L}{C_G}. \tag{3.23}$$

Very fast MOS circuits need small channel lengths L, high carrier mobilities, and small load-to-gate capacitor ratios C_L/C_G. Finally it should be mentioned that an enhancement-type load transistor can be used instead of the depletion-type transistor T_2 for the inverter of Fig. 3.2 (Carr and Mize 1972; Penney and Lau 1972). In this case a separate gate voltage supply V_{GG} is needed, or the gate of T_2 must be connected to V_{DD}.

CMOS

The basic gate (inverter) is made with an n-channel and a p-channel MOS field-effect transistor as shown in Fig. 3.4. We assume the same threshold voltages for the p- and n-channel transistors ($V_{tn} = V_{tp} = V_t$).

If the input voltage V_E is low ($V_E < V_t$) the n-channel transistor T_n is off and the p-channel transistor T_p is on because V_{SG} of the p-channel transistor is $V_{SG} = V_{DD} - V_{EL} > V_t$. Therefore T_p pulls up the output node to V_{DD} and we have a high-level value of the output voltage $V_{AH} = V_{DD}$. If the input voltage V_E is high (e.g. $V_{EH} = V_{DD}$) the n-channel transistor T_n is on ($V_{GS} = V_{EH} > V_t$) and the p-channel transistor T_p is off ($V_{SG} = V_{DD} - V_{EH} = 0$). Therefore T_n pulls down the output node to ground and we have a low-level output voltage $V_{AL} = 0$. Neither in the stationary high-level state nor in the low-level state does a current I flow. This means no d.c. power is required in the steady state. Only during the transition from high- to low-level (and vice versa) of the output voltage do we have a current (see Fig. 3.4). It should also be mentioned that for the ideal low-level and high-level voltage $V_{AL} = 0$ and $V_{AH} = V_{DD}$ no special aspect ratio α is necessary. This frees the designer from designing the transistors with restricted sizes. In most cases

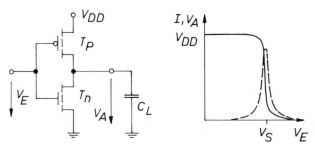

Fig. 3.4. CMOS inverter.

the transistors can be designed as small as possible, observing only the design rules.

The dynamic behaviour of the CMOS inverter is determined by the charging and discharging of the load capacitor via the transistors T_p and T_n, respectively. The time response during the fall time (discharging C_L by T_n) can be evaluated with Equation (2.265) and we obtain the fall time signal delay (see also Equation (3.6)) with $V_{EH} = V_{DD}$

$$t_{d1} = \tau_3 \frac{C_L}{C_G} \gamma \qquad (3.24)$$

where

$$\gamma = \frac{2V_t}{(V_{DD} - V_t)} + \ln \frac{3V_{DD} - 4V_t}{V_{DD}}$$

and

$$\tau_3 = \frac{L_n^2}{\mu_n (V_{DD} - V_t)}. \qquad (3.25)$$

The time response during the rise time (charging C_L by T_p) can be evaluated in the same way and we obtain, for $\mu_n \approx 2\,\mu_p$ and $V_{EL} = 0$,

$$t_{d2} = 2\alpha\tau_3 \frac{C_L}{C_G} \gamma \qquad (3.26)$$

where

$$\alpha = \frac{b_n/L_n}{b_p/L_p}.$$

In this simple estimation we have considered only one transistor (T_n or T_p) by assuming the other is off. This is not true in the transfer region from high- to low-level and vice versa (see Fig. 3.4) where both transistors are conducting. An extended analysis taking all these effects into account can be done by computer simulations (Lai 1988), but the simple formulas we derived analytically are a good first-order approximation.

BICMOS

CMOS has the advantage of low power consumption but has problems in driving large output or bus capacitors C_L in a short time. Therefore CMOS basic gates are combined with bipolar output drivers as shown in Fig. 3.5. This circuit technique is called BICMOS (Hotta *et al.* 1988; Kubo *et al.* 1988; Rosseel and Dutton 1989; Smith *et al.* 1989).

To estimate and compare the operational speed we consider the charging and discharging of a load capacitor C_L with a current I_L

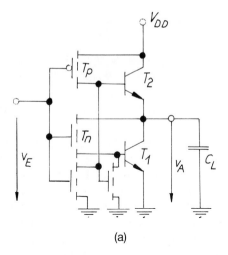

(a)

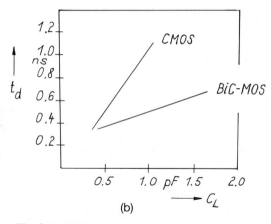

(b)

Fig. 3.5. BICMOS output driver characteristics.

$$I_L = C_L \frac{dv_A}{dt}.$$ (3.27)

If we assume a constant current I_L and define a signal delay time t_d which is needed to change the output voltage v_A on the load capacitor by ΔV (logic swing) we get

$$t_d = C_L \frac{\Delta V}{I_L}.$$ (3.28)

For pure CMOS (Fig. 3.4, and Fig. 3.5(a) without T_1 and T_2) the current I_L is the drain current, I_D, of the p-channel transistor and we have

$$t_{dCMOS} = C_L \frac{\Delta V}{I_D}. \tag{3.29}$$

In the BICMOS circuit of Fig. 3.5 the current is the emitter current of the bipolar transistors T_2 which is $I_L = I_E = B_N I_D$ and we obtain

$$t_{dBICMOS} = C_L \frac{\Delta V}{B_N I_D}. \tag{3.30}$$

The gain in operational speed for BICMOS is therefore roughly $B_N \gg 1$.

The high-level of the output voltage V_{AH} is, for the BICMOS gate in Fig. 3.5, $V_{AH} = V_{DD}$. At low input voltages V_{EL} the transistors T_p and T_2 are on and T_1 is off. For high input voltages $V_{EH} > V_t$, T_n is on and feeds a base current, I_B, into the bipolar transistor T_1. If I_B is high enough T_1 is in the saturation region and we have a low output voltage $V_{AL} \approx V_{CES}$.

Every CMOS process (n- or p-well) allows the simultaneous fabrication of bipolar transistors for BICMOS with no additional process steps.

3.1.2 Basic gates in GaAs field-effect technology

As we have seen so far the signal delay is dependent on the electron mobility, μ_n, (Equation (3.23)) and the load capacitance C_L. GaAs has a much higher electron mobility than silicon and smaller capacitances due to semi-insulation layers. Therefore GaAs circuits may have an advantage of operating speed (Ashburn et al. 1989; Shur 1987).

There are unipolar and bipolar device principles in GaAs (Shur 1987). But the most important devices currently used for GaAs circuits are the MESFET and more recently the HEMT. In the forthcoming sections we will deal with basic gates using MESFETs.

BFL

The basic gate of buffered field-effect logic (BFL) with four depletion-type MESFETs all with the same pinch-off voltage V_{pDM} is shown in Fig. 3.6. Schottky diodes are employed to provide potential shift. If the input voltage is low $V_{EL} < V_{pDM} < 0$ ($V_{pDM} < 0$, see Fig. 2.90) T_1 is off and the potential at node Z is $V_Z \approx V_{DD}$. This is high enough to drive T_4 in the conducting state and the output voltage has its high-level value $V_{AH} \approx V_{DD} - 2V_{FO} + 0.41 V_{pDM}$.

T_3 acts as a constant current source. We assume, for a simple numerical example, $V_{FO} \approx 0.5\,V$ (floating potential for all Schottky diodes, see Fig. 1.41(b)), the same pinch-off voltages $V_{pDM} = -1.0\,V$ for all MESFETs, and the same size for transistors T_3 and T_4. T_2 is much smaller than T_1.

Now we raise the input voltage to its high level V_{EH}. This switches T_1 on and the potential V_Z at node Z decreases. Therefore we now have, at the

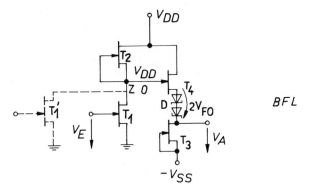

Fig. 3.6. BFL gate.

output, a low-level voltage V_{AL}. Because of the equal sizes of T_3 and T_4 and the constant current the gate–source voltages of both transistors (they are in the pinch-off region) must be zero. The low level of the output voltage is approximately $V_{AL} \approx -2V_{FO}$. For our numerical example we have $V_{AH} \approx 0$, $V_{AL} \approx -1$ V with $V_{DD} = 1.5$ V.

If we connect a second transistor T_1' in parallel with T_1 we obtain a two-input NOR gate (see Section 3.2).

SDFL

The basic gate of Schottky diode field-effect logic (SDFL) is shown in Fig. 3.7. It consists of three depletion-type MESFET and Schottky diodes. We assume for simplicity the same pinch-off voltage $V_{pDM} = -1.0$ V for all MESFETs and the same floating potential $V_{FO} = 0.5$ V for all Schottky diodes. The transistor T_1 is in the pinch-off region and acts as a constant current source.

A low level at the input (e.g. $V_{EL} \approx 0$) forces a gate potential at transistor T_2 of $V_G = -2V_{FO} = -1$ V because of the diodes and the constant

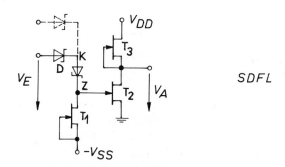

Fig. 3.7. SDFL gate.

current source. If $-2V_{FO} \le V_{pDM}$, T_2 is switched off and at the output we have a high-level voltage $V_{AH} = V_{DD}$. A high level at the input (e.g. $V_{EH} = V_{DD}$) forces a gate potential at T_2 of $V_G = V_{EH} - 2V_{FO} = V_{DD} - 2V_{FO}$. This switches transistor T_2 into the on state and the output voltage decreases down to a low-level value $V_{AL} \approx 0$ if transistor T_3 is much smaller than T_2.

Complex logic gates can be made if we apply more than one Schottky diode at the input (Vu *et al.* 1987).

DCFL

The basic gate of direct coupled field-effect logic (DCFL) is shown in Fig. 3.8. It consists of an enhancement-type MESFET (T_1, $V_{pEM} > 0$) and a depletion-type MESFET (T_2, $V_{pDM} < 0$). This gate is equivalent to the NMOS gate in Fig. 3.2 and, for the analysis of the electronic behaviour, we can apply the considerations of Section 3.1.1.

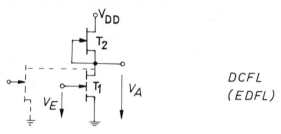

Fig. 3.8. DCFL gate.

In comparison with BFL and SDFL, DCFL has the smallest number of devices per gate but because of the simultaneous fabrication of enhancement- and depletion type MESFETs it needs a more sophisticated manufacturing process (see Section 2.3.2). Finally it should be mentioned that it is also possible to design complementary MESFET gates. (Baier 1987).

DCTL (direct coupled transistor logic) gates using HEMTs can be considered in approximately the same way as NMOS gates (see Section 3.1.1).

3.1.3 Basic gates in silicon bipolar technology

Bipolar devices have a higher gain–bandwidth product than MOS devices so they are more suitable for very high-speed circuits.

TTL

The basic gates of transistor–transistor logic (TTL) are shown in Figs. 3.9 and 3.10 (Mano 1979). The transistors are clamped with Schottky diodes to achieve small signal delays (see Section 2.1.9, Fig. 2.26).

The basic gate in Fig. 3.9 is a switch or inverter which we have already con-

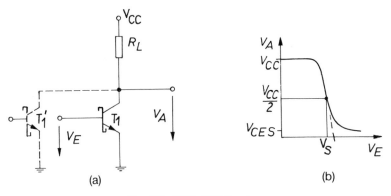

Fig. 3.9. TTL NOR gate.

sidered in Section 2.1.9, Fig. 2.23. The transfer characteristic in Fig. 3.9(b) can be derived with the simple expression for the collector current

$$I_C = I_S \exp \frac{V_E}{V_T} \tag{3.31}$$

for normal active mode of operation (see Section 2.1.2) as follows

$$V_A = V_{CC} - R_L I_S \exp \frac{V_E}{V_T}. \tag{3.32}$$

For high input voltages V_E Equation (3.32) is not correct because the transistor is then in the saturation region and the output voltage has its low-level value V_{AL} which is equal to the small collector saturation voltage $V_{AL} = V_{CES}$.

A switching or threshold point V_S in the transfer characteristic is defined as shown in Fig. 3.9(b) $(V_A(V_E = V_S) = V_{CC}/2)$. Solving Equation (3.32) yields

$$V_S = V_T \ln \left(\frac{V_{CC}}{2R_L I_S} + 1 \right). \tag{3.33}$$

These results are also valid for the basic gate in Fig. 3.10. Here the base current I_{B1} of T_1 is forced by a multiple-emitter transistor T_0.

For low input voltages (e.g. $V_{EL} = 0$) the base current of T_0 is

$$I_{BO} = \frac{V_{CC} - V_{FO} - V_{EL}}{R_B} \approx \frac{V_{CC} - V_{FO}}{R_B} \tag{3.34}$$

where V_{FO} is the floating potential of the emitter–base diode of T_0. In this case the collector current of T_0, which is the base current of T_1, is $I_{CO} = -I_{B1} \approx 0$. Therefore T_0 is in the saturation region.

Using Equation (2.21) we obtain, for $I_{CO} = 0$,

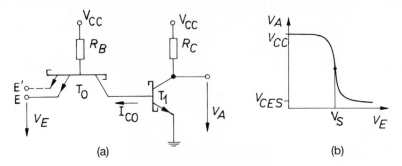

Fig. 3.10. TTL NAND gate.

$$A_N I_{ES}\left(\exp\frac{V_{FO}}{V_T} - 1\right) - I_{CS}\left(\exp\frac{V_{BCl}}{V_T} - 1\right) = 0 \tag{3.35}$$

and the base–collector voltage of T_0

$$V_{BCl} = V_T \ln\frac{I_{ES} A_N}{I_{CS}} + V_{FO} \approx V_{FO}. \tag{3.36}$$

In this case T_1 is off and the output voltage has its high-level value $V_{AH} = V_{CC}$ (see Fig. 3.10).

If we supply a high-level voltage at the input (e.g. $V_{EH} = V_{DD}$) the emitter current of T_0 is approximately zero and the collector current is (see Fig. 3.10)

$$-I_{CO} = I_{BX} = \frac{V_{CC} - 2V_{FO}}{R_B}. \tag{3.37}$$

T_0 is now in the inverse mode of operation. I_{BX} switches T_1 on and if I_{BX} is high enough T_1 will be driven into saturation. We have (see Section 2.1.9)

$$m = \frac{B_N I_{BX}}{I_{Cl}} = \frac{B_N(V_{CC} - 2V_{FO})R_C}{(V_{CC} - V_{CES})R_B} > 1 \tag{3.38}$$

and the output voltage is low: $V_{AL} = V_{CES}$ (see Fig. 3.10(b)).

As a simple numerical example, if $R_B = 5\,k\Omega$, $R_C = 700\,\Omega$, $V_{FO} = 0.7\,V$, $V_{CES} = 0.15\,V$, $B_N = 100$, and $V_{CC} = 5\,V$, we have

$$m = 100\frac{(5 - 1.4)0.7}{(5 - 0.15)5} \approx 10. \tag{3.39}$$

The dynamic behaviour of the TTL cells in Figs 3.9 and 3.10 is determined mainly by the switching behaviour of the transistor T_1. If a positive base current I_{BX} (see Equation (3.37)) switches T_1 on, the time response of the collector current may be calculated using the charge-control equation (2.129). We find

$$B_N I_{BX} = \tau_a \frac{di_{C1}}{dt} + i_{C1}. \tag{3.40}$$

With the initial condition $i_{C1}(0) = 0$, the solution of Equation (3.40) is

$$i_{C1}(t) = B_N I_{BX}\left(1 - \exp{-\frac{t}{\tau_a}}\right). \tag{3.41}$$

This current discharges the load capacitor C_L and the time response of the output voltage $v_A(t)$ can be calculated with

$$V_{CC} - R_C B_N I_{BX} = v_A + (\tau_L + \tau_a)\frac{dv_A}{dt} + \tau_L \tau_a' \frac{d^2 v_A}{dt^2} \tag{3.42}$$

where

$$\tau_a = \tau_{nB} + B_N R_C C_{sc} = \tau_a' + B_N R_C C_{sc} \tag{3.43}$$

and

$$\tau_L = R_C C_L. \tag{3.44}$$

The solution for $\tau_L \gg \tau_a$ with the initial condition $v_A(0) = V_{CC}$ is

$$v_A = V_{CC} - R_C B_N I_{BX}\left\{1 - \frac{1}{1 - \frac{\tau_a'}{\tau_L}}\exp\left(-\frac{t}{\tau_L}\right) - \frac{1}{1 - \frac{\tau_L}{\tau_a'}}\exp\left(\frac{-t}{\tau_a'}\right)\right\} \tag{3.45}$$

which is valid for $v_A > 0$. The solution for $\tau_a \gg \tau_L$ is

$$v_A = V_{CC} - R_C B_N I_{BX}\left(1 - \exp{-\frac{t}{\tau_a}}\right) \tag{3.46}$$

which is valid for $v_A > 0$.

We define a fall time signal delay (high–low delay) $v_A(t_{d1}) = V_{AH}/2 = V_{CC}/2$ (see Fig. 3.3) and derive

$$t_{d1} = \tau_a \ln\frac{m}{m - 0.5} \tag{3.47}$$

where

$$m = \frac{B_N I_{BX} R_C}{V_{CC}}.$$

If we supply, at $t = t_0$, a low input voltage $V_{EL} = 0$ the collector current of T_0 gives a negative base current of T_1

$$I_{CO} = -I_{BY} = B_N\frac{V_{CC} - V_{FO} - V_{EL}}{R_B} = B_N\frac{V_{CC} - V_{FO}}{R_B} \tag{3.48}$$

which switches the transistor T_1 off. First we have a delay time t_s (the storage time, see Section 2.1.9) where the collector current stays constant:

$$t_s = \tau_S \ln \frac{m + k}{1 + k} \tag{3.49}$$

with

$$m = \frac{B_N I_{BX} R_C}{V_{CC}}$$

and

$$k = \frac{B_N I_{BY} R_C}{V_{CC}}.$$

This delay time is due to the excess charge stored in the transistor during saturation. For Schottky TTL we do not have such a signal delay.

After the decay of the excess charge in the base, the transistor again enters, at $t_0' = t_o + t_s$, the active region, the collector current decays, and the load capacitor will be charged via the resistor R_C. If the collector current decays much faster than the capacitor can be charged via R_C we have

$$\frac{V_{CC} - v_A}{R_C} = C_L \frac{dv_A}{dt}. \tag{3.50}$$

The solution, with the initial condition $v_A (t_o') = V_{AL} = V_{CES}$, is

$$v_A(t) = (V_{CC} - V_{CES})\left(1 - \exp -\frac{t - t_o'}{\tau_L}\right) + V_{CES}. \tag{3.51}$$

The low–high signal delay time (see Fig. 3.3) with

$$v_A(t - t_o' = t_{d2}) = \frac{V_{CC}}{2} \tag{3.52}$$

is

$$t_{d2} = \tau_L \ln \left(2 + \frac{2V_{CES}}{V_{CC}}\right) \approx \tau_L \ln 2. \tag{3.53}$$

If the collector current decays slowly so that $\tau_a \gg \tau_L$ we have to insert τ_a instead of τ_L in Equations (3.51)–(3.53).

If we connect, in the basic gate of Fig. 3.9, several transistors in parallel we obtain NOR gates and if we use multi-emitter transistors in the basic gate of Fig. 3.10 we obtain NAND gates (Johnson *et al.* 1979).

ECL and CML

The basic cell of ECL (emitter-coupled logic, without emitter follower, known as current mode logic) is shown in Fig. 3.11. The operation is based

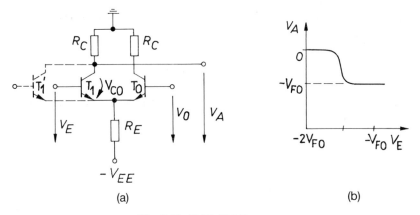

Fig. 3.11. ECL NOR gate.

on current switching. A constant current I_E ($\approx V_{EE}/R_E$) flows through a reference transistor T_0 or a logic transistor T_1 depending on the input voltage V_E. The base potential V_0 of the reference transistor is held constant. If the input voltage V_E has its low-level value V_{EL}, T_1 is off and the constant current flows through T_0. The output voltage at the collector of T_1 has its high-level value $V_{AH} = 0$. If the input voltage V_E has its high-level value V_{EH}, T_1 is on and the constant current flows through T_1 causing a voltage drop on R_C and the output voltage V_{AL} is low.

For compatibility of input and output voltage swings we must have

$$V_{EH} - V_{EL} = V_{AH} - V_{AL}. \qquad (3.54)$$

We choose an operating point of the conducting transistor with $V_{BC} = 0$ to keep the transistor out of saturation (this prevents signal delay due to the storage time in the saturation region). We choose a voltage swing

$$V_{AH} - V_{AL} = V_{FO} \qquad (3.55)$$

and, with $V_{AH} = 0$,

$$V_{AL} = -V_{FO} \qquad (3.56)$$

where V_{FO} is the floating potential of the emitter–base junctions. Further we have, with $V_{BC} = 0$

$$V_{AL} = V_{EH} = -V_{FO} \qquad (3.57)$$

and, with Equation (3.54),

$$V_{EL} = -2V_{FO}. \qquad (3.58)$$

An emitter follower (not shown in Fig. 3.11) accomplishes a potential shift V_{FO} to guarantee compatibility of input and output voltage levels. The gate potential of T_0 may be $V_0 = 1.5V_{FO}$, for example.

The transfer characteristic $V_A = f(V_E)$ can be derived as follows. The collector potential of T_1 is the output voltage

$$V_A = -R_C A_N I_{ES} \exp\frac{V_{BE1}}{V_T} \qquad (3.59)$$

(where $I_S = A_N I_{ES}$). The constant current I_E is the sum of the emitter currents of T_1 and T_0

$$I_E = I_{ES}\left(\exp\frac{V_{BE1}}{V_T} + \exp\frac{V_{BE0}}{V_T}\right). \qquad (3.60)$$

With $V_{BE0} = V_0 - (V_E - V_{BE1})$ we have

$$I_{ES}\exp\frac{V_{BE1}}{V_T} = \frac{I_E}{1 + \exp\dfrac{V_0 - V_E}{V_T}} \qquad (3.61)$$

and finally we obtain

$$V_A = \frac{-R_C A_N I_E}{1 + \exp\dfrac{V_0 - V_E}{V_T}} = -\frac{V_{FO}}{1 + \exp\dfrac{V_0 - V_E}{V_T}}. \qquad (3.62)$$

ECL gates are very fast because they are not driven into saturation and the switch-off currents are very high ($k \approx B_N V_{FO}/2r_{bb'}$ $I_E \gg 1$) (Horowitz *et al.* 1990), but, because of the constant current, the power consumption of these gates is very high.

I^2L

I^2L (integrated injection logic) gates are made with p-n-p lateral transistors as power supplies and inverse (upward) n-p-n planar transistors as switches. A cross-section of the classical I^2L structure is shown in Fig. 3.12. The n-epitaxy layer is grounded and is the emitter of all inverse n-p-n transistors and the base of the p-n-p transistors. The collectors on the top are the output terminals. The p-emitter of the p-n-p transistor is connected to the power supply V_B and is called the injector. In Fig. 3.13 a circuit model of the I^2L gate is sketched.

First, let us describe the operation qualitatively. If the input voltage V_E is low no injection current I_1 can flow into the base of the inverse (upward) n-p-n transistor and the collectors of that transistor are open (floating). If an identical gate is connected with its input to the output of that gate we have a high-level value of the output voltage V_A which is equal to the floating potential V_{FO} of the emitter–base junction (input) of the connected gate.

If the input voltage is high, $V_E \geq V_{FO}$, the injector current I_1 flows into the base of the inverse (upward) n-p-n transistor and switches it into the con-

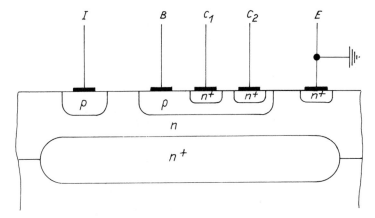

Fig. 3.12. Cross-section of an I^2L gate.

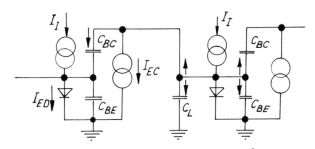

Fig. 3.13. Network model of two connected I^2L gates.

ducting state. The output voltage at the collectors decreases to a low-level value $V_{AL} = V_{CES} \approx 0$. The collector current is now due to the injector current of the connected gate. If all injector currents are equal we have

$$m = \frac{B_{NA} I_{BX}}{I_{C1}} = B_{NA} \tag{3.63}$$

where B_{NA} is the upward (inverse) current gain of the n-p-n planar transistor. To drive the gate into saturation it must be true that $B_{NA} > 1$.

The injector current can be written as (see Section 2.1.3)

$$I_I = I_{po} \exp \frac{V_B}{V_T}. \tag{3.64}$$

This is equal to the base current if the input voltage is high, $V_E = V_{EH}$ (see Fig. 3.13). Using the model of Fig. 2.16 and $B_{NAB} \gg B_{NA} > 1$ we obtain

$$I_{po} \exp \frac{V_B}{V_T} = I_{ESA}(1 - A_{NA}) \exp \frac{V_{EH}}{V_T} = \frac{I_{no}}{B_{NA}} \exp \frac{V_{EH}}{V_T} \tag{3.65}$$

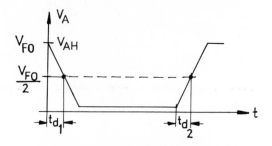

Fig. 3.14. Switching response of an I^2L gate.

where $I_{no} = A_{NA} I_{ESA}$, B_{NA} and A_{NA} are the upward (inverse) current gains, and B_{NAB} is the downward (normal) current gain of the planar n-p-n transistor.

For compatibility of the voltage levels at the input and output we must have $V_{AH} = V_{EH}$ and we obtain, with Equation (3.65), the high-level voltage

$$V_{AH} = V_B + V_T \ln \frac{I_{po}}{I_{no}} B_{NA} \approx V_B = V_{FO}. \qquad (3.66)$$

The low-level value V_{AL} is the saturation voltage V_{CES} given by Equation (2.123).

The dynamic behaviour is determined by the charging and discharging of the base–emitter and base–collector capacitors, C_{BE} and C_{BC} (see Fig. 3.13). In any case the constant injector current causes a potential swing of $V_{AH} - V_{AL} \approx V_{FO}$ at the output node and, with

$$I_1 = C \frac{dv_A}{dt}, \qquad (3.67)$$

we find a signal delay time, defined by Fig. 3.14, of

$$t_{d1} = t_{d2} = \frac{C V_{FO}}{2 I_1} \qquad (3.68)$$

where C is the effective node capacitance at the output. We have (see Fig. 3.13) $C = 3 C_{BC} + C_{BE} + C_L$. Therefore

$$t_d = \frac{(3 C_{BC} + C_{BE} + C_L)}{I_1} V_{FO}. \qquad (3.69)$$

If we use multicollector structures we can make wired OR gates (see Fig. 3.15).

There are many derivatives of this basic I^2L gate.

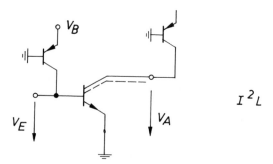

Fig. 3.15. I^2L inverter.

ISL

The basic gate of ISL (integrated Schottky logic) is shown in Fig. 3.16. A base current is provided by a voltage source V_B and a resistor R_B. The n-p-n transistor is clamped by a p-n-p substrate transistor to keep the transistor out of saturation. The voltage swing $V_{AH} - V_{AL}$ at the output is given by the difference of the floating potentials V_{FO} of the Schottky diodes at the output and the p-n junctions of the transistor, and is usually $V_{AH} - V_{AL} = 0.2\,\text{V}$.

A logic operation can be realized by a wired OR of the Schottky diodes at the output.

STL

The basic gate of STL (Schottky transistor logic) is shown in Fig. 3.17. It is similar to ISL but the transistor is clamped by a Schottky diode.

The voltage swing is given by the difference of the floating potentials of the two Schottky diodes, $V_{AH} - V_{AL} = V_{FO1} - V_{FO2}$, which is dependent on the barrier heights V_{B2} and V_{B1} ($V_{FO1} - V_{FO2} \approx V_{B1} - V_{B2}$, see Section 1.5.3). The problem here is to make Schottky diodes of quite different barrier heights on the same chip.

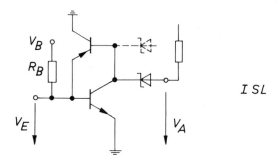

Fig. 3.16. ISL inverter.

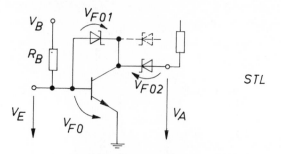

Fig. 3.17. STL inverter.

ISL and STL are very fast because of the small voltage swings $V_{AH} - V_{AL}$ and the clamped transistors. I^2L, ISL, and STL belong to the family of fan-out logic devices, whereas TTL and ECL/CML are fan-in logic devices (Johnson *et al.* 1979; Mano 1979; Shur 1987).

3.2 STANDARD CELLS AND MACROCELLS

Standard cells are small circuit units (such as logic gates and flip-flops) of which macrocells or systems are made. They can be predesigned and stored in a hardware 'library' for repeated use. Standard cells are available in all of the basic technologies we have considered in Section 3.1 (see Fig. 3.18). Macrocells are large standard cells, such as register files and arithmetic-logic units, which are made with standard cells.

In this section we will concentrate mainly on CMOS standard cells because of their great importance for future VLSI and ULSI circuits and

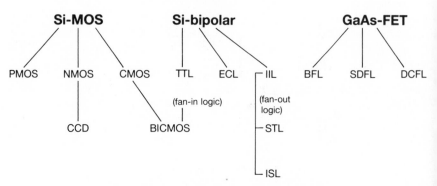

Fig. 3.18. Overview of semiconductor logic.

systems, but most of our considerations are also valid and useful for all other circuit techniques (Yuan and Svensson 1989).

3.2.1 Transfer gates

Transfer gates are the smallest CMOS standard cells. They are electronic switches consisting of a p-channel and an n-channel transistor as shown in Fig. 3.19. They connect an input node E to an output node A if the control voltage φ is high ($\varphi = 1$, $\bar{\varphi} = 0$).

1. Let us first consider the case when the input voltage is high, $V_E = V_H$, and the node capacitor C_K is charged via the transfer gate. The n- and p-channel transistors are conducting if the control voltage is also high, $\varphi = V_H$ and $\bar{\varphi} = 0$. The gate–source voltage of the n-channel transistor is $V_{GSN} = V_H - v_A(t) > V_{tn}$ and of the p-channel transistor is $-V_{GSP} = V_H > V_{tp}$. If the output voltage v_A on the output node capacitor C_K is increased to $V_{AO} = V_H - V_{tn}$ the n-channel transistor goes off and the final charging of the capacitor C_K up to V_H is done by the p-channel transistor alone.

2. Now let us consider the case when the input voltage is low, $V_E = 0$, and the node capacitor C_K is discharged via the transfer gate. Suppose, initially, that the voltage on C_K is V_H. The n-channel and p-channel transistors are conducting if the control voltage at their gates is $\varphi = V_H$ and $\bar{\varphi} = 0$. The gate–source voltage of the n-channel transistor is $V_{GSN} = V_H > V_{tn}$ and of the p-channel transistor $-V_{GSP} = V_H - v_A(t) > V_{tp}$. If the output voltage v_A has decreased to $V_{AO} = V_H - V_{tp}$, the p-channel transistor goes off and the final discharge of the capacitor C_K down to $v_A = 0$ is done by the n-channel transistor alone.

The dynamic behaviour is determined by the charging and discharging of the capacitor C_k via the n- and p-channel transistors. Using the results of Section 3.1.1. we have, for the delay times (see Fig. 3.3) of a transfer gate,

$$t_d \approx 1.5\,\tau\,\frac{C_L}{C_G} \qquad (3.70)$$

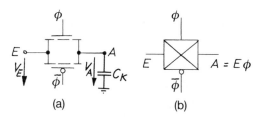

(a)   (b)

Fig. 3.19. Transfer gate: (a) CMOS circuit (b) logic symbol.

where

$$\tau = \frac{L^2}{\mu_n (V_H - V_t)} \tag{3.71}$$

and

$$C_G = \frac{\epsilon_i b L}{d_i}. \tag{3.72}$$

Fig. 3.19 depicts the transistor circuit (a) and the logic symbol (b).
    A transfer gate performs a logical AND

$$A = E \cdot \varphi. \tag{3.73}$$

This means that we have a high-level voltage at the output if, at the input E and at the control gate, the voltage is also high level.

3.2.2 Logic gates

Binary logic is based on Boolean algebra (Johnson *et al.* 1979; Mano 1979). We assume here that a high voltage level V_{EH}, V_{AH} corresponds to a logical '1' and a low-level voltage V_{EL}, V_{AL} corresponds to a logical '0'. A logical operation can be described by Boolean equations or by truth tables (Mano 1979).
    The NOT (invert) operation is

$$A = \bar{E} \tag{3.74}$$

and is implemented by an inverter such as the one shown in Fig. 3.20. This circuit has already been analysed in Section 3.1 for many circuit techniques. The truth table of the NOT operation is given in Table 3.1.
    A NOR gate (NOT OR) is shown in Fig. 3.21. The NOR operation is

$$A = \overline{E1 \vee E2} \tag{3.75}$$

and is given by the truth Table 3.2.

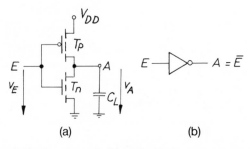

(a)   (b)

Fig. 3.20. Inverter: (a) CMOS circuit (b) logic symbol.

Table 3.1. Truth table of NOT operation.

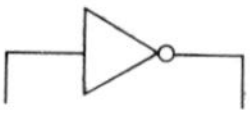

E	A
0	1
1	0

Table 3.2. Truth table of NOR operation.

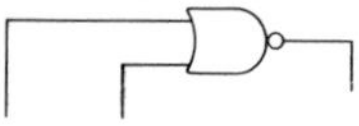

E1	E2	A
0	0	1
1	0	0
0	1	0
1	1	0

Table 3.3. Truth table of NAND operation

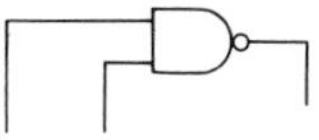

E1	E2	A
0	0	1
0	1	1
1	0	1
1	1	0

A NAND gate (NOT AND) is shown in Fig. 3.22. The NAND operation is

$$A = \overline{E1 \cdot E2} \qquad (3.76)$$

and is given by the truth Table 3.3.

An XOR gate (exclusive OR) is shown in Fig. 3.23. The XOR operation is

$$A = E1 \cdot \overline{E2} \vee \overline{E1} \cdot E2 = E1 + E2 \qquad (3.77)$$

which is given by the truth Table 3.4.

Figs 3.20–3.23 show that, for static CMOS gates, an OR/NOR operation needs n-channel transistors in parallel and corresponding p-channel transistors in series, whereas an AND/NAND operation needs n-channel transistors in series and corresponding p-channel transistors in parallel. A disadvantage of these static CMOS logic gates is that, for every input, a pair

Table 3.4 Truth table of XOR operation

E1	E2	A
0	0	0
1	0	1
0	1	1
1	1	0

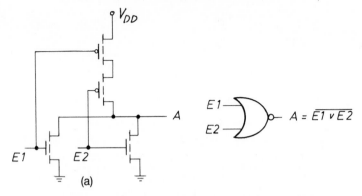

Fig. 3.21. NOR gate: (a) CMOS circuit (b) logic symbol.

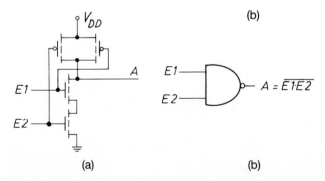

Fig. 3.22. NAND gate: (a) CMOS circuit (b) logic symbol.

of n- and p-channel transistors is needed. This can be avoided by the domino logic shown in Fig. 3.24. The logic network of domino logic consists only of n-channel transistors in series and/or parallel as required to realize the logical operation. This network is pulled up to the power supply V_{DD} by a p-channel transistor T_p and pulled down to ground by an n-channel transistor T_n. If the clock voltage φ is low, the p-channel transistor is on and

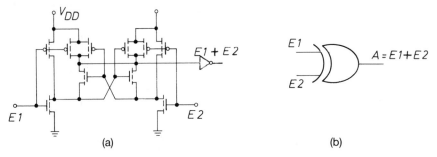

Fig. 3.23. XOR gate: (a) CMOS circuit (b) logic symbol.

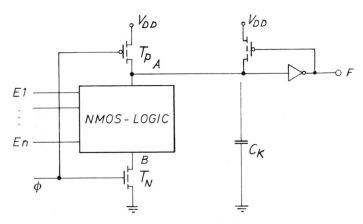

Fig. 3.24. Domino logic.

the n-channel transistor is off. Therefore the output node A is pulled up to V_{DD} (C_K will be charged). Then the clock voltage φ increases and the n-channel transistor is on, whereas the p-channel transistor is off. Now the output node A will be discharged ($A = 0$) or remains high ($A = 1$) depending on the logical operation being performed by the network of the n-channel transistors. Depending on the logical operation, the n-channel network either forms a conducting path between A and B, or it does not. It should be noted that only in the high state of the clock voltage φ is the logic level at A is valid (0 or 1).

Domino logic is a dynamic CMOS logic. To avoid an unwanted decay of the high-level voltage at the output because of leakage currents, a circuit at the output which clamps the high-level voltage is also implemented in the gate of Fig. 3.24. (Hwang and Fisher 1989).

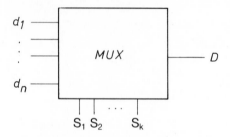

Fig. 3.25. Block symbol of a multiplexer.

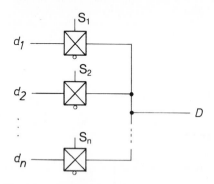

Fig. 3.26. Multiplexer with transfer gates.

3.2.3 Multiplexers

Multiplexers are macrocells that select input signals, d_1, ..., d_n, and transmit them to a single output, D, depending on a control signal vector, S_1, ..., S_k (see Fig. 3.25). Therefore a multiplexer can be understood as a logic network (Mano 1979) but in CMOS technology, multiplexers are made simply with transfer gates as shown in Fig. 3.26. If the control signal S_i is high ('1'), the data signal input d_i is connected to the output D. Multiplexers are used in complex systems to interconnect several functional blocks via common buses (see Chapter 4).

3.2.4 Decoders

Decoders and encoders are macrocells that convert an input bit pattern, a_1, ..., a_n, to an output bit pattern, f_1, ..., f_m (see Fig. 3.27). Special decoders convert an input bit pattern a_1, ..., a_n to a unique uncoded output signal, $f_i = 1$, so that an input bit pattern activates only one output signal. This is demonstrated for a simple example in Table 3.5.

From this table we derive the Boolean equations of the output signals

Fig. 3.27. Block symbol of a decoder.

Table 3.5. Truth table of a decoder.

a	b	f_1	f_2	f_3	f_4
1	1	1	0	0	0
0	1	0	1	0	0
1	0	0	0	1	0
0	0	0	0	0	1

$$f_1 = a \cdot b = \overline{\overline{a} \vee \overline{b}} \qquad (3.78\text{a})$$

$$f_2 = \overline{a} \cdot b = \overline{a \vee \overline{b}} \qquad (3.78\text{b})$$

$$f_3 = a \cdot \overline{b} = \overline{\overline{a} \vee b} \qquad (3.78\text{c})$$

$$f_4 = \overline{a} \cdot \overline{b} = \overline{a \vee b}. \qquad (3.78\text{d})$$

This decoder can be implemented with NOR gates as shown in Fig. 3.28.

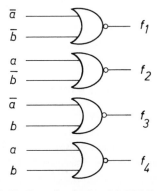

Fig. 3.28. Decoder logic with NOR gates.

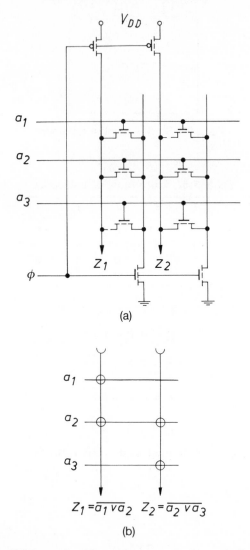

Fig. 3.29. NOR matrix: (a) transistor matrix (b) symbolic notation.

3.2.5 Matrix structures (NOR and ROM)

These macrocells are made with arrays of MOS transistors as shown in Fig. 3.29. The MOS transistors are arranged in rows and columns like a matrix which can be programmed by making some transistors active, and others not; inactive transistors are disconnected from the column or row lines (see Fig. 3.29(a)). In a simplified diagram we use circles for active transistors (fully connected to row and column lines) sketched in Fig. 3.29(b).

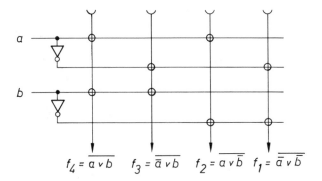

$$f_4 = \overline{a \vee b} \quad f_3 = \overline{\overline{a} \vee b} \quad f_2 = \overline{a \vee \overline{b}} \quad f_1 = \overline{\overline{a} \vee \overline{b}}$$

Fig. 3.30. Example of a decoder with NOR matrix.

With these matrices we may obtain programmable NOR operations. Our simple example of Fig. 3.29 gives the two NOR operations

$$Z_1 = \overline{a_1 \vee a_2} \tag{3.79}$$

$$Z_2 = \overline{a_3 \vee a_2} \tag{3.80}$$

as can be easily seen from the illustration. The decoder of Fig. 3.28 in Section 3.2.4 can be made with the NOR matrix of Fig. 3.30.

The advantage of these matrix arrays is that they can be predesigned and programmed (customized) at the last moment but it needs a large area of silicon and is slower than random logic because of the long column and row lines. From a memory point of view they are read-only memories (ROM). In a ROM a row corresponds to a word address (word line) and the column lines are the bit lines. For further details see Section 4.5.3.

3.2.6 PLA, FPLA, and PAL

PLA (programmable logic arrays) are macrocells which can be hardware-programmed to perform any logical operation in sum-of-product form (Mano 1979). In CMOS they are implemented by two NOR matrices (see Section 3.2.5) as shown in Fig. 3.31. The first matrix (on the left in Fig. 3.31) is called an AND plane where product terms, $P_1, \ldots, P_z$, (elementary conjunctions) of the input signals, $A_1, \ldots, A_n$, are generated. In our simple example of Fig. 3.31 these are:

$$P_1 = A_1 \overline{A}_2 \overline{A}_3 A_4$$
$$P_2 = \overline{A}_1 \overline{A}_2 A_4$$
$$P_3 = \overline{A}_1 \overline{A}_3 \overline{A}_4.$$

These product terms act as inputs into the second NOR matrix (on the right

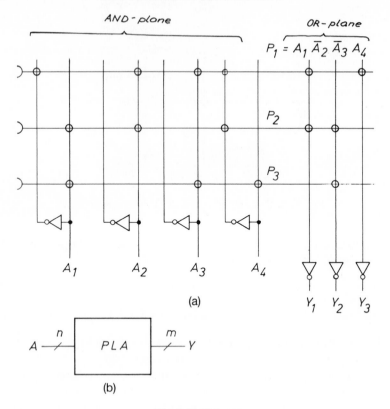

(a)

(b)

Fig. 3.31. PLA.

in Fig. 3.31) called an OR plane where the product terms are ORed to give the output signals. In our example these are:

$$Y_1 = \bar{A}_1 \bar{A}_2 A_4 \vee A_1 \bar{A}_2 \bar{A}_3 A_4 = P_1 \vee P_2 \tag{3.81}$$

$$Y_2 = \bar{A}_1 \bar{A}_2 A_4 \vee \bar{A}_1 \bar{A}_3 \bar{A}_4 = P_2 \vee P_3 \tag{3.82}$$

$$Y_3 = A_1 \bar{A}_2 \bar{A}_3 A_4. \tag{3.83}$$

Let us now consider the programming in greater detail. As mentioned in Section 3.2.5, the hardware programming of a NOR matrix is done by setting an active MOS transistor (shown by a circle in the matrix). This transistor must be set in the AND plane at the matrix cross-point of the product term and the inverted input signal line. An example may clarify this procedure. We consider P_2 of Fig. 3.31. Suppose we have to realize $P_2 = \bar{A}_1 \bar{A}_2 A_4$. Then active transistors (circles) have to be set at the cross-points P_2/A_1, P_2/A_2, and $P_2/\bar{A}_4$, as already shown in Fig. 3.31. A_3 is not involved in P_2 (i.e. 'don't care' condition).

For the notation of AND plane programming we use the following convention:

1 means a transistor is active at a non-inverted input;
0 means a transistor is active at an inverted input;
– means no transistor is active (don't care).

For our example in Fig. 3.31 we then have

$$0 \quad 1 \quad 1 \quad 0$$
$$1 \quad 1 \quad - \quad 0$$
$$1 \quad - \quad 1 \quad 1$$

For the notation of OR plane programming we use the following convention:

1 means the product term (elementary conjuction) is involved in the output operation;
0 means the product term (elementary conjunction) is not involved in the output operation;
– means it does not matter whether the product term is involved or not.

For the example of Fig. 3.31 we have the following notation of the OR plane:

$$1 \quad 0 \quad 1$$
$$1 \quad 1 \quad 0$$
$$0 \quad 1 \quad 0$$

With this formal notation we can program a computer with a description of the logical operation and the geometrical structure of a PLA. A special computer program can minimize and fold the PLA to achieve minimal silicon usage for a given logical operation. Logic minimization reduces the number of product-term lines. Folding reduces the column lines. In a folded PLA the input and output lines (which run in columns) are applied on the top and on the bottom of the matrices.

The advantage of a PLA is its flexibility in logic design. They can be predesigned and hardware-programmed at the last moment. Changes in the logic design can be carried out very easily without influencing the geometry of the neighbours on the chip but it needs a large chip area, and the long product-term lines cause a large signal delay.

FPLAs (field-programmable logic arrays) contain a number of OR and AND gates that can be interconnected (hardware-programmed) by links formed with integrated circuit fuses which can be selectively blown to break the connection. PAL (programmable array logic) also belongs to the group of fuse-linked programmable arrays. It is a collection of predesigned logic gates which are connected by a system of lines which are fixed in place. The user's choice is to break or not break a given line via a fusible link.

For FPLA and PAL the user cannot reconfigure the logic after hardware-programming through the fusible links. Mask-programmable PAL is called HAL (high-volume array logic) and PAL which can be reconfigured electrically is called GAL (generic array logic).

3.2.7 Barrel shifters

A barrel shifter is a macrocell that contains a matrix of transfer gates as shown in Fig. 3.32. It performs very effectively a multiple bit shift which is necessary for bit-field extraction and many other bit-field manipulations.

The basic topology of the barrel shifter in Fig. 3.32 makes it possible for any bus bit D_i to be available at any output position D_i'. Therefore a multiple bit shifter is no more difficult to design than a single bit shifter. The crossbar switches (transfer gates) are controlled by a control signal vector, $S_0, \ldots, S_k$. To explain the basic operation we consider the simple 4-bit example of Fig. 3.32. If the control signal S_0 is active ($S_0 = 1$) we have no shift operation and $D_1' = D_1$, $D_2' = D_2$, $D_3' = D_3$, and $D_4' = D_4$. If the control signal S_1 is active ($S_1 = 1$) a 1-bit rotate-left operation is carried out. This means $D_1' = D_4$, $D_2' = D_1$, $D_3' = D_2$, and $D_4' = D_3$. If $S_2 = 1$ we have $D_1' = D_3$, $D_2' = D_4$, $D_3' = D_1$, and $D_4' = D_2$ etc.

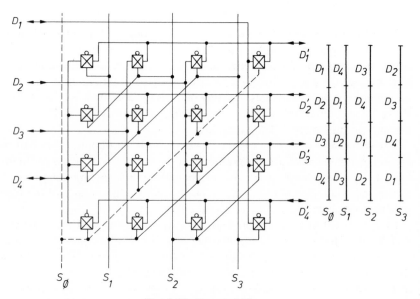

Fig. 3.32. Barrel shifter.

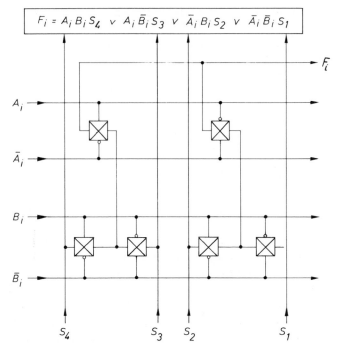

$$F_i = A_i B_i S_4 \quad \vee \quad A_i \bar{B}_i S_3 \quad \vee \quad \bar{A}_i B_i S_2 \quad \vee \quad \bar{A}_i \bar{B}_i S_1$$

Fig. 3.33. Universal logic unit with transfer gates.

3.2.8 Arithmetic-logic units

Arithmetic-logic units are the heart of every data and signal processor. They can perform logic and arithmetic operations. First, we will consider a universal logic (UL) unit which is made with transfer gates shown in Fig. 3.33. The logical input signals are A_i and B_i and the control signals are S_1, S_2, S_3, and S_4. From Fig. 3.33 we can read the logical operation

Table 3.6. Logic operations of the UL in Fig. 3.33.

S_4	S_3	S_2	S_1	F_i	name
0	0	0	1	$\overline{A_i \vee B_i}$	NOR
0	0	1	1	$\overline{A_i}$	NOT
0	1	0	1	$\overline{B_i}$	NOT
0	1	1	0	$A_i \overline{B_i} \vee \overline{A_i} B_i$	XOR
0	1	1	1	$\overline{A_i B_i}$	NAND
1	0	0	0	$A_i B_i$	AND
1	0	0	1	$A_i B_i \vee \overline{A_i} \overline{B_i}$	$\overline{\text{XOR}}$
1	0	1	0	B_i	
1	1	1	0	$A_i \vee B_i$	OR

Table 3.7 Truth table of full adder operation.

A_i	B_i	C_i	S_i	C_{i+1}
0	0	0	0	0
1	0	0	1	0
0	1	0	1	0
1	1	0	0	1
0	0	1	1	0
1	0	1	0	1
0	1	1	0	1
1	1	1	1	1

$$F_i = A_i\,B_i\,S_4 \vee A_i\,\bar{B}_i\,S_3 \vee \bar{A}_i\,B_i\,S_2 \vee \bar{A}_i\,\bar{B}_i\,S_1 \tag{3.84}$$

and we have, according to the control signal pattern $S_4\,S_3\,S_2\,S_1$, the logical operations at the output as listed in Table 3.6.

The basic elements of all arithmetic-logic units is the full adder because all other arithmetic operations (subtraction, multiplication, and division) can be done by using full adders (Mano 1979). Full adders add two binary digits, A_i and B_i, and a carry, C_i, to a sum S_i and a new carry, C_{i+1}, to the next higher binary digit. The truth Table 3.7 gives the logical operation of a full adder.

The logical operation of the sum is

$$S_i = A_i + B_i + C_i \tag{3.85}$$

and of the carry is

$$C_{i+1} = A_i \cdot B_i \vee C_i (A_i + B_i). \tag{3.86}$$

The logic circuit of a full adder bit slice is shown in Fig. 3.34. For a 32-bit adder, 32 such full adders must be connected in a chain. Every full adder bit slice can carry out the addition if the predecessor has calculated its carry C_{i+1} but this takes a certain delay time t_i. The most significant bit slice has to wait at least $32 \times t_i$ until it has its incoming carry and can perform the

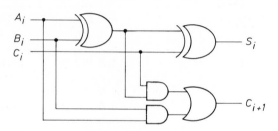

Fig. 3.34. Full adder.

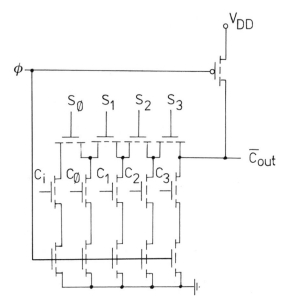

Fig. 3.35. Carry look-ahead circuit.

addition. This slows down the data processing so special carry look-ahead circuits are employed to speed the process.

In Fig. 3.35 a MOS carry look-ahead circuit is depicted. The partial sums S_i and partial carries C_i are generated from half adders (not given in the figure)

$$S_i' = A_i + B_i \qquad (3.87)$$

$$C_i' = A_i \cdot B_i. \qquad (3.88)$$

The carry C_4 is generated immediately by the chain in Fig. 3.35

$$C_4 = C_3' \vee S_3' \left(C_2' \vee S_2' \left(C_1' \vee S_1' \left(C_0' \vee S_0' \, C_i\right)\right)\right). \qquad (3.89)$$

Another solution of this problem is shown in Fig. 3.36. It is a carry bypass/generate circuit. Here we have generate signals, G_i, and propagate

Table 3.8. Genesis of the next carrier by generate and propagate signals.

C_i	G_i	P_i	C_{i+1}
x	0	0	0
x	1	0	1
1	0	1	1
0	0	1	0

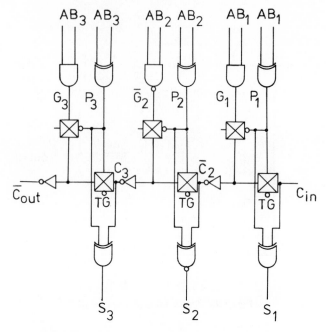

Fig. 3.36. Carry look-ahead carry bypass circuit.

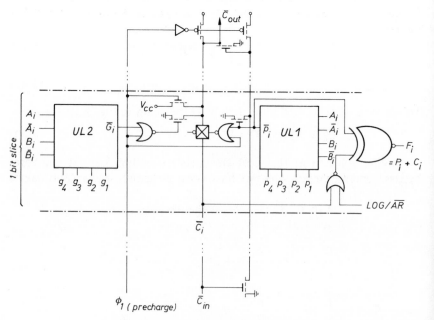

Fig. 3.37. ALU for VLSI processors.

Table 3.9. Logical operations of the ALU in Fig. 3.37.

LOG/AR	p_4	p_3	p_2	p_1	$\overline{P_i}$	F_i
1	1	0	0	1	$\overline{XOR}$	XOR
1	1	1	1	0	OR	NOR
1	0	0	0	1	NOR	OR
1	1	0	0	0	AND	NAND
1	0	1	1	1	NAND	AND
1	0	0	1	1	$\overline{A_i}$	A_i
1	0	1	0	1	$\overline{B_i}$	B_i
1	1	1	0	0	A_i	$\overline{A_i}$
1	1	0	1	0	B_i	$\overline{B_i}$

signals, P_i, dependent on the binary digits A_i and B_i. The genesis of the next carrier is given in Table 3.8. The carry is passed by transfer gates which are controlled by the propagate signal P_i and this principle is employed in the ALU of Fig. 3.37.

The control signal LOG/$\overline{AR}$ switches the ALU from arithmetic to logical operations. If LOG/$\overline{AR}$ = 1 the output is

$$F_i = P_i \qquad (3.90)$$

and the carry C_i will be ignored. In that case the universal logic unit UL1 can generate all sixteen logical operations of two binary digits, A_i and B_i, as given in Table 3.6. This depends on the control signal vector $p_4 p_3 p_2 p_1$. We repeat some of them in Table 3.9.

If LOG/$\overline{AR}$ = 0 the ALU performs arithmetic operations. The output is then

$$F_i = P_i + C_i. \qquad (3.91)$$

For addition we must have $P_i = A_i + B_i$ to perform the logical operation of a full adder (see Equation (3.85))

$$F_i = A_i + B_i + C_i. \qquad (3.92)$$

A carry must be generated if $A_i B_i = 1$ so the generate operation is

$$G_i = A_i \cdot B_i. \qquad (3.93)$$

The control signal vector for UL2 is then $g_4 g_3 g_2 g_1 = 0111$. An existing carry can pass if $A_i + B_i = 1$ so the propagate operation is

$$P_i = A_i + B_i. \qquad (3.94)$$

The control signal vector for UL1 is then $p_4 p_3 p_2 p_1 = 1001$. This is summarized in Table 3.10.

Subtraction is performed by an addition with the 2s complement of the

Table 3.10 Arithmetic operations (A) of the ALU in Fig. 3.37.

LOG/$\overline{\text{AR}}$	p_4	p_3	p_2	p_1	g_4	g_3	g_2	g_1	$\overline{P_i}$	$\overline{G_i}$	C_{in}	F_i
0	1	0	0	1	0	1	1	1	XOR	A_iB_i	0	A_i plus B_i
0	1	0	0	1	0	1	1	1	XOR	A_iB_i	1	A_i plus B_i plus 1

Table 3.11 Arithmetic operations (B) of the ALU in Fig. 3.37.

LOG/$\overline{\text{AR}}$	p_4	p_3	p_2	p_1	g_4	g_3	g_2	g_1	$\overline{P_i}$	$\overline{G_i}$	C_{in}	F_i
0	0	1	1	0	1	0	1	1	XOR	$\overline{A_i\bar{B}_i}$	1	A_i minus B_i
0	0	1	1	0	1	0	1	1	XOR	$\overline{A_i\bar{B}_i}$	0	A_i minus B_i minus 1

subtrahend ($\bar{B}_i + 1$) and we have the operations P_i and G_i as given in Table 3.11.

Multiplication can be carried out by repeated addition with shifted multiplier (pencil and paper method). The circuit diagram of a serial–parallel multiplier is shown in Fig. 3.38. For the multiplication of two binary numbers of n digits ($A_1 \ldots A_n$ and $B_1 \ldots B_n$) we need two shift registers (see Section 3.2.10) and n full adders. In Section 4.3.5 we will give another implementation of a multiplier.

Division can be done by repeated subtraction of the shifted divisor.

As we have seen so far combinational logic circuits have a wide range of

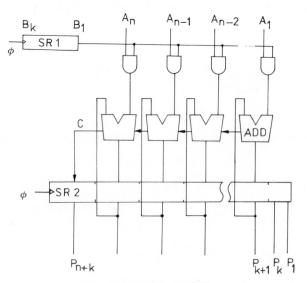

Fig. 3.38. Serial multiplier.

applications from extremely simple to highly complex but their output at any instant is determined only by the inputs at that instant. In the next sections we will deal with sequential circuits.

3.2.9 Flip-flops

Flip-flops are the basic elements of sequential circuts. Sequential circuits generate output that depends on present as well as past inputs, they can remember or store information from the previous inputs, and their logical operations F depend on the input signal E and the internal state Z $F = f(E, Z)$.

The classical flip-flop is the RS flip-flop shown in Fig. 3.39; it can be made from two NOR or NAND gates. The input signals are R (reset, clear) and S (set, preset). If $S = 1$ the output state Q is set to $Q = 1$ ($\bar{Q} = 0$) and if $R = 1$ the output state Q is reset to $Q = 0$ ($\bar{Q} = 1$). The operation is given in the transition Table 3.12. (The flip-flop in Fig. 3.39 is clocked which means the inputs, R and S, are active only if the clock pulse φ is high.)

We obtain a D flip-flop (data latch) if S is connected to the D input and R is connected to the inverted D input as shown in Fig. 3.40. The operation of the D flip-flop is given in Table 3.13.

A JK flip-flop is obtained by connecting to the R and S inputs signals given by the logic equations (Fig. 3.41)

$$R = Q \cdot K \tag{3.95}$$

$$S = \bar{Q} \cdot J. \tag{3.96}$$

The transition table is given in Table 3.14. In particular, for $JK = 1$ the next state, Q_{n+1}, is the inversion of the previous state Q_n. This can be used to build frequency dividers (toggle flip-flops, see Section 3.2.11).

Table 3.12. Transition table of the RS flip-flop.

R	S	Q_{n+1}
0	0	Q_n
1	0	0
0	1	1
1	1	x (indeterminate)

Table 3.13. Transition table of the D flip-flop.

D	Q_{n+1}
0	0
1	1

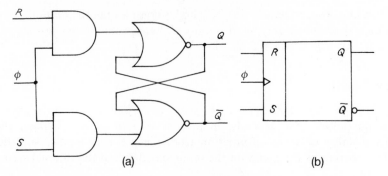

Fig. 3.39. RS flip-flop.

Table 3.14. Transition table of the JK flip-flop.

J	K	Q_{n+1}
0	0	Q_n
0	1	0
1	0	1
1	1	$\overline{Q}_n$

In synchronous systems the flip-flops are gated with a clock pulse or are edge-triggered (Mano 1979). In the next sections when we consider more complex sequential standard cells and macrocells we will also deal with such gated and edge-triggered flip-flops.

3.2.10. Register files

Memory registers

Memory registers are used for data storage. They are made with latches (D flip-flops) of Fig. 3.40 or Fig. 3.42. The latch in Fig. 3.42 is a semistatic latch made with transfer gates and inverters. This latch is gated with a two-phase non-overlapping clock with voltages φ_1, φ_2. If $\varphi_1 = 1$ ($\varphi_2 = 0$) the data signal D is coupled via a transfer gate to the first inverter and is stored temporarily on the node capacitor C_k. At $\varphi_2 = 1$ ($\varphi_1 = 0$) a feedback loop is closed via a transfer gate and the data signal is stored permanently in the latch. If we use, instead of the second inverter, a NOR gate, a reset signal R can be applied to this latch.

The implementation of a register file using such latches is shown in Fig. 3.43. A 32-bit register file with N registers contains $32 \times N$ latch cells. The bit lines of the register file are coupled via tristate drivers (see Fig. 3.56)

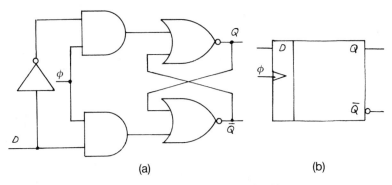

Fig. 3.40. *D* flip-flop (data latch).

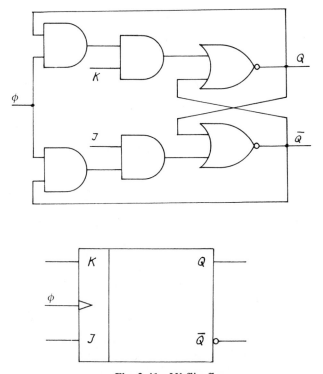

Fig. 3.41. *JK* flip-flop.

to a data bus. Register selection is possible with read and write address lines RA_n and WA_n, respectively.

For a write operation to the nth register the WRITE and the write-address select signals, WA_n, must be active ($= 1$). During clock phase $\varphi_1 = 1$ ($\varphi_2 = 0$) the data signal on the bus are coupled and stored in the selected

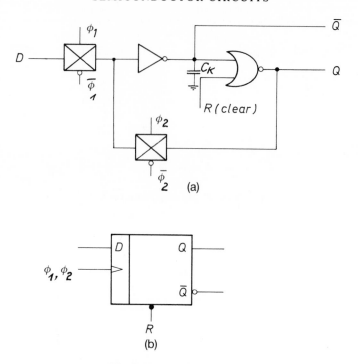

Fig. 3.42. Semistatic latch.

latches. During the clock phase $\varphi_2 = 1$ ($\varphi_1 = 0$) the data signal is latched steadily.

For a read operation of the nth register the READ and the read-address select signal RA_n must be active (= 1). During clock phase $\varphi_1 = 1$ the data signal is read out of the latch on to the data bus.

This is an example of a register file working with a single bus (see Fig. 3.44(a)) but for fast and parallel data processing, register files with two and three buses are better (see Fig. 3.44(b), (c)). As an example we show in Fig. 3.45 a register file with three data buses A, B, and C. As a latch, we use here a static CMOS flip-flop. Now we have two read-address select signals RA_n and RB_n and one write-address select signal W_n for every register. In this example we can only write from the bus C and read on to the buses A and B. It should be mentioned that for a write operation of the static CMOS flip-flop in Fig. 3.45 the inverted and the non-inverted data signals, B_i and $\bar{B}_i$, must be available.

Register files are used in processor chips as general data and address registers, status registers, and as tables and pointers (Horowitz 1990).

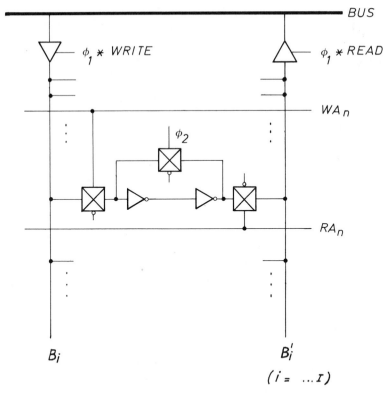

Fig. 3.43. Register file serving a single bus.

Shift registers

A shift register can store data like a memory register and, additionally, it can move data. It can shift data from one stage of the register to an adjacent stage from right to left or vice versa and data may be loaded in and read out of the register in serial or in parallel. This gives very interesting options for data manipulation and transmission (Ashburn et al. 1989; Mano 1979).

The classical standard cell for shift registers is the master–slave flip-flop shown in Fig. 3.46(a). A semistatic master–slave flip-flop is shown in Fig. 3.46(b). This is implemented with three transfer gates and two inverters. The data signal DI_i is loaded via a transfer gate into the first inverter and is stored temporarily on node 1 if $\varphi_1 = 1$ ($\varphi_2 = 0$, $\varphi_2' = 0$). If $\varphi_2 = 1$ ($\varphi_1 = 0$) it is transmitted to the adjacent inverter and a short time later (if $\varphi_2' = 1$) a feedback loop is closed. Now the data signal is permanently stored in the shift register cell and is available at the output DO_i.

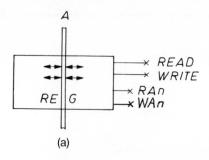

(a)

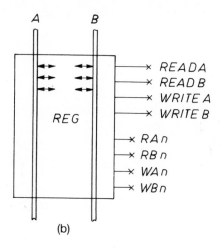

(b)

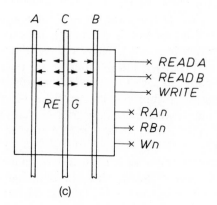

(c)

Fig. 3.44. Principles of register files serving one, two, or three buses.

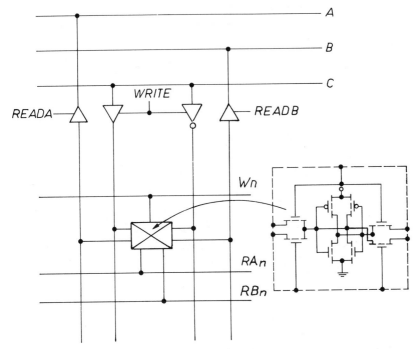

Fig. 3.45. Register file serving three buses.

The full dynamic shift register cell of Fig. 3.47 has no feedback path. Therefore the data signal can only be stored temporarily on the node capacitors, 1 and 2, and must be kept moving otherwise leakage currents will discharge (or charge) the nodes 1 and 2 and will destroy the data information.

Let us look at an example. If the node capacitor (on node 1 or node 2) is $C_k = 0.1$ pF and the leakage current is $I = 10^{-10}$ A the time during which we have a voltage swing of $\Delta V = 1$ V on node 1 or 2 is

$$t = \frac{C_K}{I} \Delta V = 1\,\text{ms}. \qquad (3.97)$$

This means the lowest possible clock frequency for that shift register will be 1 kHz.

The shift register chain of Fig. 3.48 performs a first-in-first-out operation (FIFO). The data which is loaded first into the shift register at the serial input DI_i comes out first at the serial output DO_i. A shift operation during phase 1 is enabled by activating the LOAD signal.

The shift register of Fig. 3.49 performs the last-in-first-out operation (LIFO). If the control signal PUSH is active (PUSH = 1) a data bit is loaded via the transfer gate at the input during clock phase φ_1 and latched

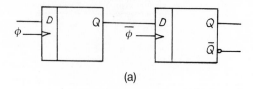

(a)

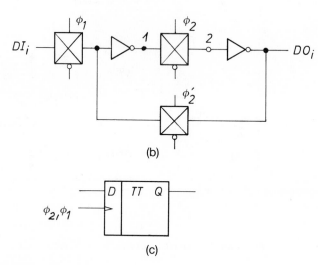

(b)

(c)

Fig. 3.46. Master–slave flip-flop: (a) logical principle (b) semistatic cell (c) logic symbol.

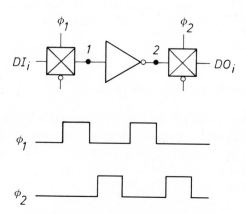

Fig. 3.47. Dynamic shift register cell.

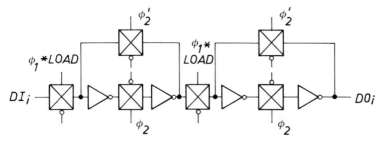

Fig. 3.48. Semistatic shift register cell for FIFO.

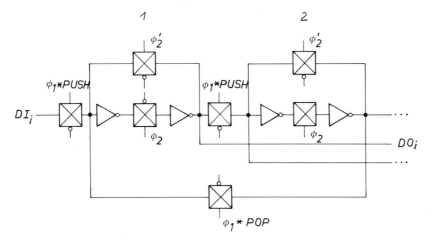

Fig. 3.49. Semistatic shift register cell for LIFO.

during phase φ_2'. All other bits in the shift register chain are pushed one position forward (to the right in Fig. 3.49). If the control signal POP is active (POP = 1) all data bits are moved (shifted) during phase φ_1 one position backward (to the left in Fig. 3.49) and at the output DO_i we have the last-pushed data bit. Such LIFO registers are applied as stacks.

3.2.11 Counters and timers

There are many principles for counters and timers. A counter counts the number of input pulses it receives and stores them. The simplest design is the ripple counter. The standard cell of that type is the toggle flip-flop (T flip-flop) which can be made with a JK flip-flop where the J- and K-inputs are connected together (see Table 3.14).

In Fig. 3.50 we show an alternative semistatic design using transfer gates and inverters as is usual in CMOS VLSI. In Fig. 3.50(c) we show the timing signals at the input T, the output Q, and the internal nodes X and Y. If we

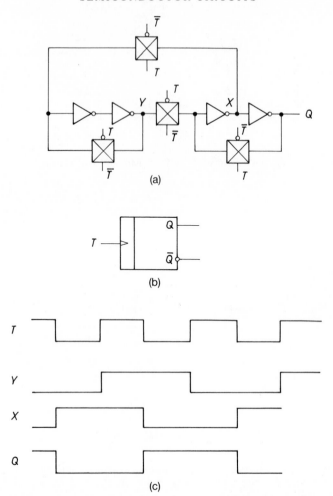

Fig. 3.50. T flip-flop: (a) semistatic cell (b) logic symbol (c) timing diagram.

compare T and Q we see that Q has half the frequency of T; this is a frequency divider. If we connect n such T flip-flops in a chain we obtain a binary ripple counter counting from 0 to 2^{n-1}.

A three-stage ripple counter is shown in Fig. 3.51 where the output bits are $Q_A Q_B Q_C$. If we force a reset at a certain output bit pattern $Q_A Q_B Q_C$ we can implement any counting limit. To clarify this we design a three-stage ripple counter counting from 0 to 5 so that the reset signal must be generated at the output bit pattern $Q_A Q_B Q_C = 101$. The logic operation for the reset signal is therefore

$$R = Q_A \bar{Q}_B Q_C. \tag{3.98}$$

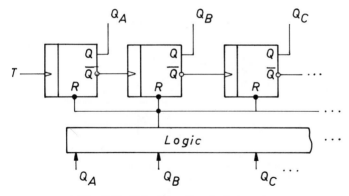

Fig. 3.51. Principle of a ripple counter.

Ripple counters have the disadvantage that the count pulse must ripple through all n stages of the counter in series. If the signal delay time of a single stage is t_d the overall signal delay of the ripple counter is at least $n \times t_d$.

Synchronous counters are designed to prevent this long delay. In synchronous counters the clock pulse is connected to all stages simultaneously and we use JK flip-flops instead of T flip-flops. Each of these flip-flops is therefore triggered at the same time. The J- and K-inputs are logical operations of the outputs Q_A, Q_B, ... of all stages, and the different logical operations which are connected to the J- and K-inputs of every individual JK flip-flop depend on the counting scheme. An example is given in Fig. 3.52. It is a BCD (binary coded decimal) counter which counts from 0000 to 1001 (0, ..., 9). Synchronous counters are faster and less susceptible to errors arising from unwanted or shortened pulses caused by the cumulative delay of the asynchronous ripple counters.

Counters with arbitrary and programmable counting schemes are employed as timers for the control of digital systems. Such timers are made

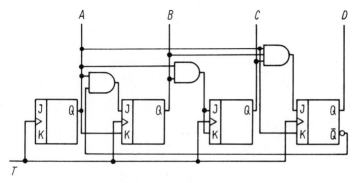

Fig. 3.52. Synchronous BCD counter.

$$DI = (\bar{Q}_A \vee Q_C)\bar{Q}_B$$

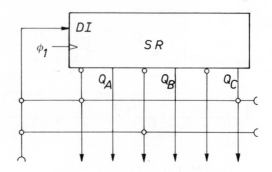

Fig. 3.53. Shift register counter.

with shift registers or PLA. In Fig. 3.53 a counter/timer with shift register and PLA is shown. Its operation can be explained as follows. In the NOR matrix (AND plane) two product terms are generated (see also Section 3.2.6)

$$P_1 = Q_A \bar{Q}_C \tag{3.99}$$

$$P_2 = Q_B. \tag{3.100}$$

These are NORed to give the input signal of the shift register

$$DI = \overline{P_1 \vee P_2} = (\bar{Q}_A \vee Q_C)\bar{Q}_B \tag{3.101}$$

So that we have the counting sequence

$$\begin{array}{ccc} 0 & 0 & 0 \\ 1 & 0 & 0 \\ 0 & 1 & 0 \\ 0 & 0 & 1 \\ 1 & 0 & 0 \end{array}$$

etc.

The principle and an example of a counter/timer with PLA is shown in Fig. 3.54. The output bit pattern is $Q_A Q_B Q_C$ and the pattern of the internal state is $Y_1 Y_2$ (state vector). The state vector is generated in the OR plane and is transmitted via the state register into the AND plane. For this example we have the logical operations of the output bit pattern

$$Q_A = \bar{A}_1 \vee A_1 \bar{A}_2 \tag{3.102}$$

$$Q_B = A_1 A_2 \vee A_1 \bar{A}_2 = A_1 \tag{3.103}$$

$$Q_C = A_1 \bar{A}_2 \tag{3.104}$$

MOORE AUTOMATON

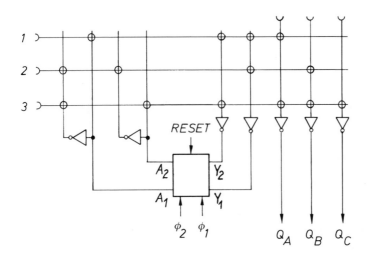

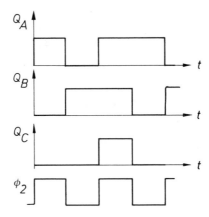

Fig. 3.54. An example of a PLA counter.

and the state vector

$$Y_1 = \bar{A}_1 \vee A_1 A_2 \tag{3.105}$$

$$Y_2 = \bar{A}_1 \vee A_1 \bar{A}_2 \tag{3.106}$$

(to verify this, see Section 3.2.6). After reset ($Y_1 Y_2 = 00$) the automaton steps through the states 00 11 10 01 11 and the sequence of the output bit pattern is, for $Q_A Q_B Q_C$,

$$1 \quad 0 \quad 0$$
$$0 \quad 1 \quad 0$$
$$1 \quad 1 \quad 1$$
$$1 \quad 0 \quad 0$$
$$0 \quad 1 \quad 0$$

etc. This type of automaton is called a Moore automaton because the output bit pattern is dependent only on the internal states of the system.

3.2.12 I/O circuits

At the input of every I C we need special circuits which

(1) adjust the voltage and current levels (e.g. from TTL to MOS); and
(2) protect the circuit from distortions.

In Fig. 3.55 we show a typical input circuit for CMOS ICs. The resistor R (5 kΩ) together with the MOS diode T protects the gates of the CMOS inverter from high voltages. A punch-through of T at V_{pt} (see Section 2.2.4, Fig. 2.54) causes a current which is limited by the resistor R. T is designed for a punch-through voltage V_{pt} which is much smaller than the gate breakdown voltage $V_{BR} = d_i E_{crit}$ ($E_{crit} = 5 \times 10^6$ V/cm).

At the output of internal functional blocks we need driver circuits which are able to

(1) drive the large bus or output capacitors in a short time; and
(2) switch their output in a tristate.

In Fig.3.56 we show a CMOS tristate driver circuit. The p-and n-channel transistors T_p and T_n are big enough to drive the PAD or bus capacitors with the required speed. If the chip enable signal CE is low ($CE = 0$, $\overline{CE}$

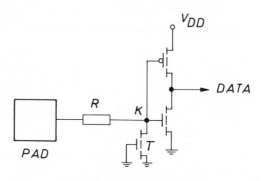

Fig. 3.55. CMOS input circuit.

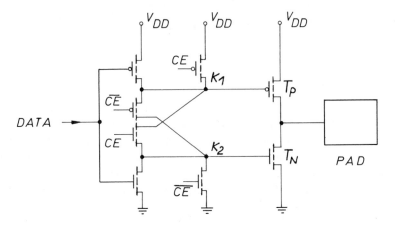

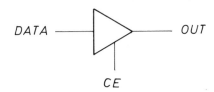

Fig. 3.56. CMOS output driver.

= 1) the gate potential of T_p is V_{DD} and of T_n is 0 so T_p and T_n are off. This makes the output PAD floating (tristate).

If the chip enable signal CE is high ($CE = 1$, $\overline{CE} = 0$) the output is controlled by the input $DATA$ signal. In modern very fast CMOS ICs we have BICMOS drivers as previously shown in Fig. 3.5.

3.3. PRINCIPLES OF SIMULATION

After every design step it is very important to verify the operation by simulation. A hierarchy of simulation methods and tools from system simulation to layout verification (see also Fig. 4.4) is needed. For standard cell and macrocell circuit and logic design the following methods are applied:

(1) network simulation;
(2) timing simulation;
(3) switch-level simulation;
(4) logic gate simulation.

We will explain briefly what they are and how they work in principle but we will not describe the details of commercially available simulation tools such as SPICE and HILO etc.

3.3.1 Network simulation

Network simulation is the most sophisticated and detailed method for functional verification of transistor networks (Ohtsuki 1986a). We need here a non-linear network model of the circuit and we will demonstrate this for the simple CMOS circuit in Fig. 3.57. The non-linear network models of transistors are given in Sections 2.1.6 and 2.2.5. In particular, for the circuit of Fig. 3.57 we have the voltage-dependent current sources $i_1, \ldots, i_4$ as shown in Fig. 3.58. The voltage dependence of the transistor currents i_k is given by the model equations Equations (2.219)–(2.226). The node capacitors, C_2 and C_3, and the feedback capacitor C_{23}, are also voltage dependent (see Equations (2.227)–(2.235)) Furthermore in the circuit we have control functions, $F_1(t)$ and $F_2(t)$, as input signals (see Fig. 3.59) and node voltages, $v_2(t)$ and $v_3(t)$, as unknown variables.

From Kirchhoff's law we obtain an implicit non-linear system of differential equations for the calculation of the vector of all node voltages $\mathbf{Y} = v_1 v_2 v_3 \ldots$

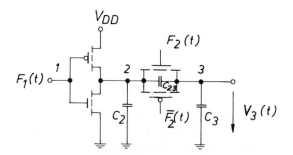

Fig. 3.57. CMOS circuit with transfer gate.

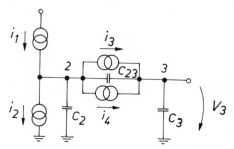

Fig. 3.58. Network model of the CMOS circuit of Fig. 3.57.

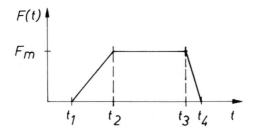

Fig. 3.59. Input function.

$$F(\mathbf{Y}, \mathbf{\mathring{Y}}, t) = 0. \tag{3.107}$$

This system will be solved numerically by a network analysis program such as SPICE. Modern network analysis programs employ highly sophisticated and stable implicit integration algorithms, and for fast solution of the large system of linear equations sparse matrix techniques are applied (Ohtsuki 1986a). Such network analysis programs are time-consuming and reasonable only for networks having no more than 500 nodes. Several attempts have been made to shorten the run time and to save memory space (Ohtsuki 1986a).

3.3.2 Timing simulation

Timing simulation is a simplified network simulation which uses simple integration formulas and simple transistor models. The node voltage VK at the instant t will be calculated with

$$VK(t) = VK(t - 1) + \frac{I}{CK} S(t - 1) \frac{\text{DELTA}}{\text{DUKMAX}(t - 1)} \tag{3.108}$$

where

$$\text{DUKMAX}(t) = \text{MAX}(VK(t) - VK(t - 1)). \tag{3.109}$$

If DUKMAX is larger than a given DELTA then the time step S is corrected by

$$S(t) = \frac{\text{DELTA}.S(t - 1)}{\text{DUKMAX}(t)}. \tag{3.110}$$

CK is the node capacitor and I is the resulting current at a node. It is the sum of all transistor currents which flow to or from the node. A simple transistor current model for timing simulation is (see Equations (2.220) and (2.221) with K4 = 0, K5 = 1), for $VGS - Vt > VDS$,

$$I = K(2(VGS - Vt)VDS - VDS \uparrow 2) \tag{3.111}$$

and, for $VGS - Vt \le VDS$,

$$I = K(VGS - Vt) \uparrow 2 \qquad (3.112)$$

$$Vt = VtO + K2\sqrt{VSB}. \qquad (3.113)$$

Timing simulation needs less simulation time and memory space than network simulation but it sometimes causes stability problems.

3.3.3 Switch-level simulation

It is not the aim of switch-level simulation to calculate the time response of the node voltages VK(t). In the simplest case it calculates the discrete logic states 0 and 1 at the nodes so that switch-level simulation is a logic simulation which considers all transistors as switches. As an example we will explain the principle by the following algorithm:

First a bit pattern is supplied to the inputs and all internal nodes are filed in an actual node list. Then the states of all nodes available in this node list are determined. We have four different states:

state 1: isolated node;

state 2: the node will be charged (it has only a conducting path to power supply V_{DD});

state 3: the node will be discharged (it has only a conducting path to ground);

state 4: The node will be charged and discharged (it has conducting paths to ground and power supply).

A conducting path is searched by considering all gate potentials of the transistors in the path. If the gate potential is high then the transistor contributes to a conducting path; if not then the path is interrupted by this transistor.

Now we can assign the logic levels to all nodes in the different states. If the node has

state 1   the logic level is kept unchanged;

state 2   the logic level is set to 1;

state 3   the logic level is set to 0;

state 4   the logic level is set to 1 if the path to the power supply is more conducting than the path to ground, otherwise it is set to 0.

Now all nodes which have not changed will be deleted in the actual node list. Then all nodes which are influenced by the nodes currently in the actual node list will be searched and all nodes which have been found by these path search mechanisms will be put into the actual node list. This will be repeated until the actual node list is empty which means all nodes have their stable logic level.

This method is an event-driven simulation which considers only those circuit parts where something is happening. In very complex digital systems such as microprocessors, this gives reasonably low simulation times. In such systems usually only a small part of the whole transistor network is busy if an information processing task (e.g. a instruction execution) is carried out. This algorithm is especially suitable for logic simulation of MOS/CMOS networks. Because of the many circuit design tricks (dynamic circuits, transfer gate logic) the logic network is transistor (switch) oriented rather than logic gate oriented.

3.3.4. Logic gate simulation

If the system or functional block is described by a network of standard logic gates and flip-flops or by Boolean equations a logic gate simulation may be preferred. Models for every standard logic gate and flip-flop, describing the logical operation and the signal delay, are implemented in the simulation program. The network of logic gates is then described by a special language, e.g.

$$G1 = XOR; 5; G17; G10$$

This means the output of gate G1 is an XOR operation of the output of gate G17 and G10 which takes a delay time of 5 ns etc. ....

The simulation starts with an input test pattern which will be processed step by step in all gates involved in discrete time steps. Modern logic gate simulation programs are also event driven. The result is a sequence of logic levels (1, 0, or indeterminate) at every discrete time step at all nodes (outputs of the gates).

3.4 PROBLEMS FOR CHAPTER 3

1. Derive and plot the transfer characteristic of the inverter with ohmic load (Fig. 3.60) by using the simple I/V relations for the MOS transistors in Equations (2.166) and (2.167).
2. Derive and plot the transfer characteristic of the inverter with enhancement-type load (Fig. 3.61) by using the simple I/V relations for MOS transistors in Equations (2.166) and (2.167).
3. Derive and plot the transfer characteristic of the inverter with depletion-type load (Fig. 3.62) by using the simple I/V relations for the MOS transistors in Equations (2.166) and (2.167).

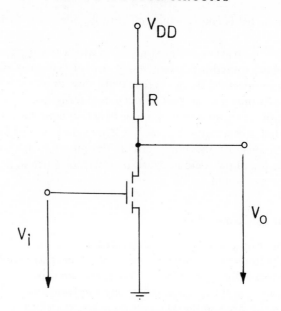

Fig. 3.60. MOS inverter with resistive load.

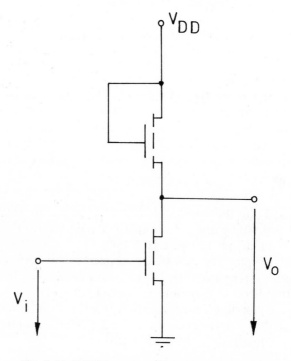

Fig. 3.61. MOS inverter with enhancement load.

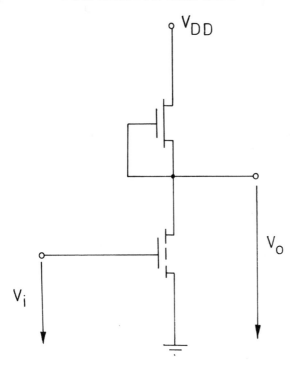

Fig. 3.62. MOS inverter with depletion load.

4. Derive and plot the transfer characteristic and the current of a CMOS inverter by using the ideal I/V relations for MOS transistors with $K_n = 2K_p$. (Hint: Observe that for p-channel transistors all signs of currents and voltages have to be reversed in Equations (2.166) and (2.167). Use the same absolute value for n-channel and p-channel threshold voltages $|V_{tp}| = |V_{tn}|$.)

5. For TTL compatibility the transfer characteristic has to fit into the scheme shown in Fig. 3.63. An NMOS inverter with depletion load (see Fig. 3.62) has to convert TTL into MOS voltage levels at the input of a MOS circuit. The threshold voltage for the enhancement–type and depletion–type transistors are $V_{tE} = + 0.8$ V and $-V_{tD} = 3$ V, respectively. Calculate the ratio of width to length for both transistors (aspect ratio $\alpha = (b_E/L_E)/(b_D/L_D)$).

6. A tristate output driver consists of two n-channel enhancement-type MOS transistors T_1 and T_2 as shown in Fig. 3.64.
   (a) Derive the high-level value of the output voltage V_{OH} if $V_2 = V_{DD} = 5$ V and $V_1 = 0$.
   (b) Calculate the low-level value of the output voltage V_{OL} if

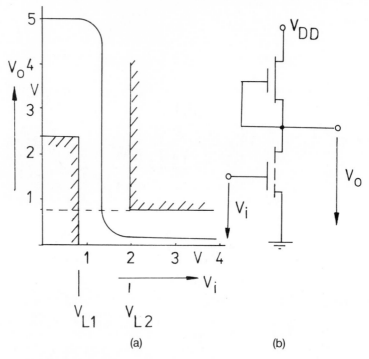

Fig. 3.63. (a) TTL output characteristic (b) MOS input circuit.

$V_1 = V_{DD}$, $V_2 = 0$ and the load current (e.g. coming from a TTL gate as shown in Fig. 3.64) is $I_o = 1$ mA.
The following parameters are given: $V_t = V_{to} + K_2\sqrt{V_{SB}}$; $V_{to} = 1$ V; $K_2 = 0.5$ volt; and $K_1 = K_2 = 5$ mA/V^2.

7. Find a CMOS circuit with a minimum number of transistors which has the logical function (XOR)

$$f = a\bar{b} \vee \bar{a}b.$$

8. Consider the differential amplifier with n-channel enhancement–type MOS transistors in Fig. 3.65. All transistors operate in pinch-off and have the current–voltage relation

$$I = K(V_{GS} - V_{tE})^2$$

where $V_{tE} = 1$ V. Find the transistor model constants K for all transistors.

9. A chain of inverters is given in Fig. 3.66. All transistors have the ideal current–voltage characteristics according to Equations (2.166) and (2.167). The threshold voltage for the n-channel enhancement-type transistors are $V_{tE} = V_{DD}/5$ and for the depletion-type transistors $-V_{tD} =$

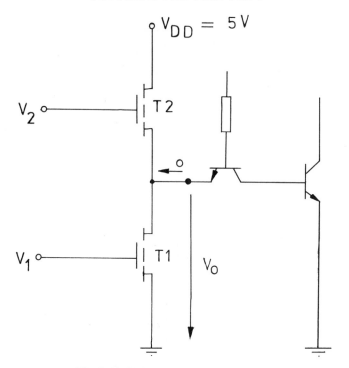

Fig. 3.64. MOS–TTL interface circuit.

3 V. The transistor current constant K for the depletion-type transistors is $K_D = 0.020\ mA/V^2$. The chain is driven by a low-going input pulse V_{in} as shown in Fig. 3.66 and $C_L = 0.05pF$

(a) Calculate the turn-on delay t_{on} for T_3 when T_3 is turned on at $v_o(t_{on}) = 2V_{tE}$ and the initial value for v_o is $v_o(0) = 0.5$ V.

(b) What is the maximum voltage at node C_L?

10. What is the logic function of the circuit shown in Fig. 3.67? Find the logic network with NOR gates which has the same logic function.

11. Design a logic network for a comparator of two nibbles. The output function should be '1' if the two nibbles are equal.

12. Design a logic network with NOR gates for a decoder which has the following truth table

F_1	F_2	F_3	F_4	D_1	D_2
1	0	0	0	0	0
0	1	0	0	0	1
0	0	1	0	1	0
0	0	0	1	1	1

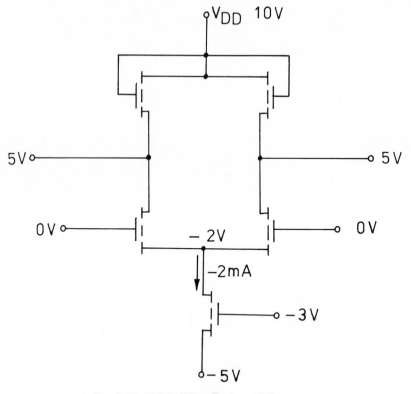

Fig. 3.65. MOS differential amplifier stage.

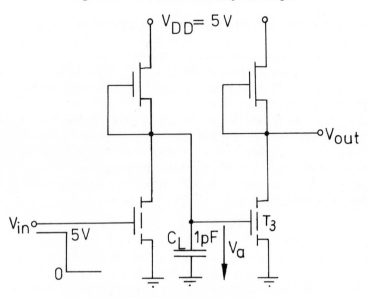

Fig. 3.66. MOS inverter chain.

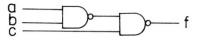

Fig. 3.67. Logic circuit.

13. Decimal numbers can be coded in BCD as shown in the following table

Decimal digit	BCD code
0	0 0 0 0
1	0 0 0 1
2	0 0 1 0
3	0 0 1 1
4	0 1 0 0
5	0 1 0 1
6	0 1 1 0
7	0 1 1 1
8	1 0 0 0
9	1 0 0 1

pseudotetrades ≥ 1 0 1 0

If we add two BCD digits it is possible that the result is greater than 1001 in which case we have to correct these 'pseudotetrades' by addition of 6 (0110) and generate a decimal carry in the next higher decimal digit.
   Draw the logic block diagram of an adder for two BCD digits to a sum BCD digit and a (optional) decimal carry in the next higher BCD digit. Use, for the solution, two functional blocks each with 4-bit full adders.

14. The truth table for a BCD to 7-segment decoder is given below

D	C	B	A	a	b	c	d	e	f	g
0	0	0	0	1	1	1	1	1	1	0
0	0	0	1	0	1	1	0	0	0	0
0	0	1	0	1	1	0	1	1	0	1
0	0	1	1	1	1	1	1	0	0	1
0	1	0	0	0	1	1	0	0	1	1
0	1	0	1	1	0	1	1	0	1	1
0	1	1	0	1	0	1	1	1	1	1
0	1	1	1	1	1	1	0	0	0	0
1	0	0	0	1	1	1	1	1	1	1
1	0	0	1	1	1	1	1	0	1	1

(a) Find the Boolean equations for a, b, c, d, e, f, g.
(b) Draw a logic diagram using NOR and NAND gates.

15. Find the logic network which acts as a demultiplexer (Fig. 3.68) with the following truth table

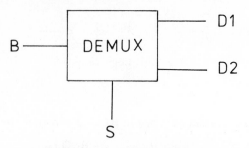

Fig. 3.68. Block symbol of a multiplexer.

S	B	D_1	D_2
0	0/1	B	0
1	0/1	0	B

16. Define a PLA with three inputs and the following output functions

$$Y_1 = \bar{A}_1 \, \bar{A}_2 \, A_3 \vee A_1 \, A_2$$
$$Y_2 = A_1 \, \bar{A}_2 \, \bar{A}_3 \vee A_1 \, A_2 \, A_3$$
$$Y_3 = \bar{A}_1 \, A_2 \, \bar{A}_3$$
$$Y_4 = A_1 \, A_2 \, \bar{A}_3 \vee A_1 \, \bar{A}_2 \vee A_3.$$

17. Design a MOS network of a simple arithmetic-logic unit (ALU) with a minimum number of transistors which performs the following functions

Control signals		Carry	Result (function)
S_2	S_1	C	F
0	0	0	A plus B
0	0	1	A plus B plus 1
0	1	0	A minus B minus 1
0	1	1	A minus B
1	0	x	A (NOT A)
1	1	x	AND

18. An 8-bit ALU has 8 output lines $D_7, \ldots, D_0$ for the result bits and 1 line, C_{out}, for carry. Derive from $D_7, \ldots, D_0$ and C_{out} the following FLAGS
    S = sign
    Z = zero
    C = carry
    V = overflow (which is set in sign 2 s complement representation when the result is greater than $+ 127$ or less than $- 128$).

19. Sketch the logic diagram of a clocked JK flip-flop with provision of SET and RESET (asynchronous).

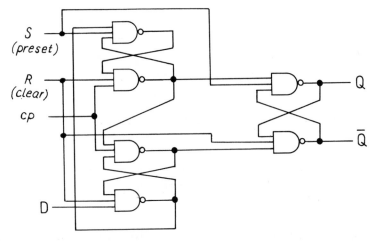

Fig. 3.69. Logic circuit of an edge-triggered D flip-flop.

20. Obtain the logic diagram of a D master–slave flip-flop using NOR gates.

21. In Fig. 3.69 an edge-triggered D flip-flop using NAND gates is shown. Convince yourself that the output is only influenced during the low–high edge of the clock pulse, cp.

22. Design a four-stage ripple counter which counts from 0 to 13.

23. Design a three-stage synchronous binary counter using JK flip-flops. It counts from 000 to 111.

24. Design a BCD counter using a JK flip-flop as
    (a) a ripple counter, and
    (b) a synchronous counter.

25. Design a programmable 3-bit counter with PLA. It has one input control signal B. The counting sequences are:

    for $B = 0$ $A = 000$
    　　　　　　　　001
    　　　　　　　　010
    　　　　　　　　011
    　　　　　　　　000 etc.

    for $B = 1$ $A = 000$
    　　　　　　　　001
    　　　　　　　　110
    　　　　　　　　111
    　　　　　　　　000 etc.

    Hint: First draw a flow diagram for the algorithm.

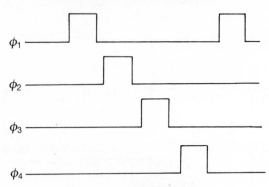

Fig. 3.70. Four-phase clock.

26. A four-phase clock in fig. 3.70 is to be generated by a shift register and logic. Find the circuit.
27. Sketch a single-shot circuit using signal delay in gates.

4 VLSI systems

4.1 INTRODUCTION

4.1.1 The evolution of VLSI systems

Recent advances in semiconductor technology have made it possible to design and fabricate chips with hundreds of thousands or millions of transistors operating at clock speeds higher than 20 MHz while the minimum feature size has decreased dramatically (Fig. 4.1). An integrated-circuit chip containing such a large number of components is really a large electronic system rather than a circuit (Mead and Conway 1980). These systems can contain various processors, memory systems, I/O parts, data converters, and analogue parts.

The hierarchy of a VLSI system is shown in Fig. 4.2. The system consists of megacells which comprise several thousand transistors and are made up of functional blocks (macrocells) such as register files and airthmetic-logic units (Hennesy 1984; Mukherjee 1985). The macrocells are made with standard or leaf cells which, in turn, usually contain just a few transistors. The lowest level in the hierarchy is the layout level or physical device level.

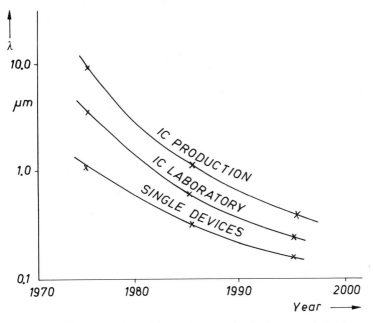

Fig. 4.1. History of the minimum feature size in integrated circuits.

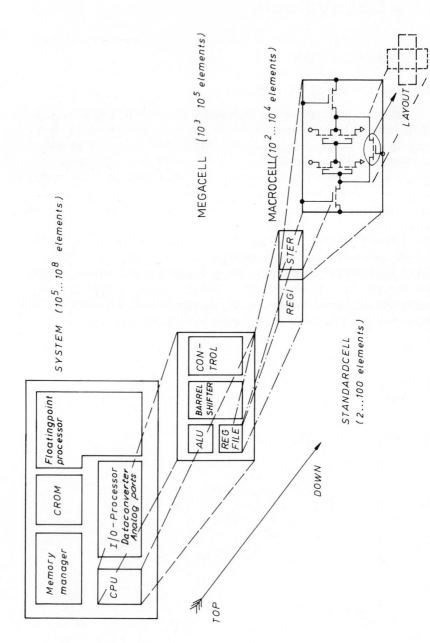

Fig. 4.2. Hierarchy of a VLSI chip.

4.1.2 Design style

Because of the enormous amount of data in the design process of a VLSI chip the design is carried out from the highest level to the lowest level in the hierarchy of Figs. 4.2 (Clements 1985; Mead and Conway 1980). This top-down design style is shown in Fig. 4.3 and 4.4 and is summarized in Table 4.1. Corresponding to this hierarchy we have several levels of design and description.

1. At the systems level the architecture of the system will be fixed. In particular, for digital processors this is the definition of data types, instructions,

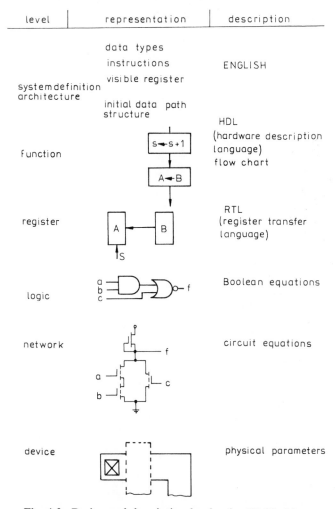

Fig. 4.3. Design and description levels of a VLSI chip.

Table 4.1. Hierarchy of VLSI design and description levels.

LEVEL	OBJECTS	LANGUAGE	TOOL	RESULT
SYSTEM SPECIFICATION ARCHITECTURE	data types instructions visible registers initial data path structure	English	human brain	users manual
ALGORITHM	statements flowchart boxes algorithmic constructs	hardware description languages (HDL) programming languages (e.g. PASCAL)	compiler interpreter debugger system simulator	functional algorithm
FUNCTIONAL REGISTER	registers functional blocks buses final data path structure	register transfer languages (RTL)	library of functional blocks and related clusters of micro-operations programs for data & control path synthesis RT–simulator	fixed data path & string of microoperations fixed control path

LOGIC	logic gates PLA ROM	Boolean equations truth tables state tables	programs for logic synthesis microprogram assembler & allocator logic & switch-level simulator	microcode PLA definition random logic
NETWORK	transistors	network description language (NBS)	network & timing simulator	electrical transistor network
LAYOUT	polygons	graphic description (CIF)	graphics editor sticks placement & routing programs verification programs	design file

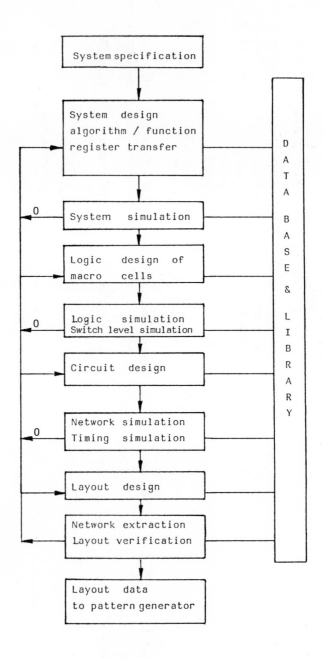

Fig. 4.4. Top-down design.

visible registers flags, interrupts, I/O signals, P A Ds, and an initial data path structure. The ingenuity of the design engineer is the most important factor here.

2. At the algorithmic level a functional algorithm of the system is worked out and is described by a flowchart and/or a higher programming or hardware description language (H D L).

3. At the register level the hardware structure of the system (data and control path) including functional blocks and buses will be designed. This can be described by a register transfer language (R T L). Associated with the fixed data path is a string of micro-operations and, for the design of the final data path, parallel and pipeline operations have to be included to fulfil the timing demands.

4. The logic design of the functional blocks includes:
   (a) synthesis of random logic;
   (b) programming of P L As and the microcode definition of matrix structures (R O Ms).
This design is described in terms of Boolean equations and state diagrams, as well as truth, state, and transition tables.

5. At the network level the transistor circuits are designed and described by a network description language.

6. The lowest level is the physical layout level. This will be treated in Chapter 5.

Every design step needs to be verified by simulation or comparison as shown in Fig. 4.4.

4.2 DATA TYPES AND INSTRUCTION FORMATS

Information processing usually employs digital techniques (Denyer and Renshaw 1985; Khung 1988) so we confine our attention to digital systems only. We obtain the digital electrical signals from the non-electric analogue signals of the real world by means of sensors and data converters (Gray 1989). Digital binary signals are strings of 0s and 1s, and such strings can represent binary data (numbers, characters) or instructions (Mano 1982; Wilkinson 1986). A binary digit is called a bit, 8 bits make up a byte, 2 bytes a word, etc. A bit string of N bits can code up to 2^N different items of data or instructions (e.g. using bytes we can code $2^8 = 256$ different numbers or characters). In Fig. 4.5 we show some important data formats as they are processed in V L S I systems.

Single bytes are used mostly as character codes (e.g. A S C I I code) or as two nibbles of packed decimal code (e.g. BCD) (Mano 1982; Wilkinson 1986). Words, longwords, and quadwords are used as signed or unsigned

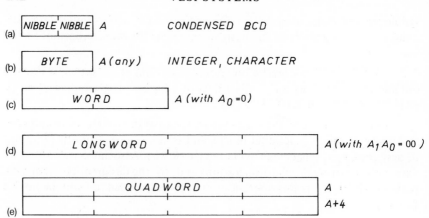

Fig. 4.5. Integer data formats.

integer numbers. The most significant bit (msb) is the sign bit which, if msb = 1, the number is negative in 2s complement representation, and if msb = 0, the number is positive.

It is also possible to use words, longwords, and quadwords as fixed-point numbers, e.g.

$$1011.0110 = 1 \times 2^3 + 0 \times 2^2 + 1 \times 2^1 + 1 \times 2^0 + 0 \times 2^{-1} + 1 \times 2^{-2}$$
$$+ 1 \times 2^{-3} + 0 \times 2^{-4} = 11.375$$

Floating-point numbers are represented by a bit string for the mantissa and a bit string for the exponent as shown in Fig. 4.6. The arithmetic operations with integer, fixed- and floating-point numbers are treated elsewhere (Mano 1982; Wilkinson 1986) and will not be repeated here.

Fig. 4.6. Floating-point data format.

The application of bit strings for instruction codes is shown in Fig. 4.7. The instruction format consists of three main parts:

(1) operation code (opcode);

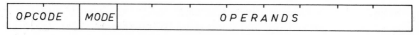

Fig. 4.7. Instruction format.

(2) addressing modes (mode);

(3) operands.

The opcode is assigned for a macro-instruction. Some typical macro-instructions and their mnemonics are given in Table 4.2.

There are many addressing modes some of which will be reviewed briefly here.

Immediate   The operand is specified in the instruction by an integer number (literals). For example,

MOVB R1, $C8

means that the byte $C8 is moved into the register R1.

Register direct   Here a register is specified holding the operand. For example,

ADDW R3, R4

means the word available in register R4 is added to the word in register R3 and the result is stored in register R3.

Table 4.2. Macro-instructions.

mnemonic	operation
MOVE	move data
STO	store data
IN	input data
OUT	output data
PUSH	push data on to stack
POP	pop data from stack
ADD	add data
SUB	subtract data
AND	logical and
OR	logical or
XOR	exclusive or
RR	rotate right
SL	shift left
BR	branch
BZ	branch if zero
BC	branch if carry
JMP	jump
CALL	call a subroutine
RET	return from subroutine

Register indirect (register deferred)   Here the content of a register specified in brackets is the operand address in memory. For example,

MOVL R2, (R4)

means that the longword located in the memory at an address which is stored in register R4 is moved into register R2.

Absolute (extended)   Here the memory address of an operand is specified in the instruction. For example,

MOVL R1, LABEL

means the operand which is available in the memory location LABEL is moved into register R1.

Memory indirect (absolute deferred)   The address of the operand is given as a memory address in bracket. For example,

MOVB R2, @ (LOCA)

means that the byte whose address is stored in memory at the location LOCA is moved into register R2. This operation is illustrated in Fig. 4.8 for a numerical example in which LOCA = $00CC12AA.

Relative   Here is a displacement relative to the instruction pointer (program counter, PC) or some other base address is given as operand. For example

BRC d

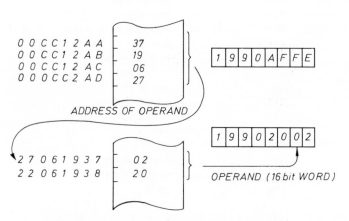

Fig. 4.8. An example of an instruction execution.

RISC INSTRUCTION FORMAT

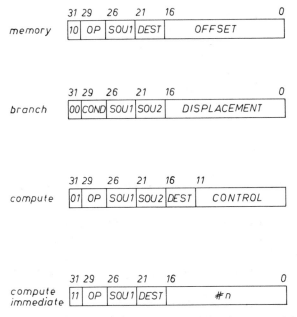

Fig. 4.9. RISC instruction format.

means the instruction pointer is added with an 8-, 16-, or 32-bit displacement if the carry is set. This results in a jump to location P C + d in the instruction code memory.

There are many more addressing modes for complex instruction set computer (CISC) processors if we use some kind of indexing, scaled indexing, base addressing, and post- and pre-increment and decrement (Crawford 1986; MacGregor *et al* 1984; Supnik 1984). On the other hand for reduced instruction set computer (RISC) processors we have a small number of instructions available all of which have the same length (e.g. 4 bytes). In Fig. 4.9 we give an example of instruction formats for a RISC processor. The instructions are classified by the most significant bits as memory, branch, and compute instructions. The opcodes (OP) or the branch condition (COND) are specified with 3 bits. The data source (SOU) and destination register (DEST) are specified next. Finally some special information on offset, displacement, literals, and control are given.

4.3 DATA-PATH ARCHITECTURES

A digital system can be subdivided into two main parts, the execution unit (information processing unit) and the control unit (Mano 1982; Tredennick

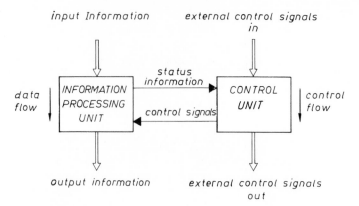

Fig. 4.10. Data and control flow in a processor chip.

1988). This is shown in Fig. 4.10. The task of the execution unit is to perform logical operations, arithmetic computations, and bit-field manipulations. It tells the control unit something about the result, e.g. whether it is zero or negative, or whether an overflow has occurred, etc. Sometimes this status information is called a flag. The control unit supplies micro-operation signals (control signals) to step the execution unit through all the micro-operations needed to execute a macro-operation. It is a finite-state machine (see Section 4.4) which is controlled by external instruction signals as well as internal status signals (flags).

4.3.1 Bit-serial and bit-parallel architectures

Basically we distinguish between bit-serial and bit-parallel data-path architectures (Denyer and Renshaw 1985). We will demonstrate these principles with a simple example, the addition of two nibbles of integer data $A = A_3 \ldots A_0$ and $B_0 \ldots B_3$, in Figs 4.11 and 4.12.

In parallel data-path architectures (Fig. 4.11) all bits of the operand are supplied at the same instant to the execution unit (four full adders) and all bits of the result are issued in parallel at the output. In bit-serial data-path architectures (Fig. 4.12) only one bit is processed (added in our example) at a time (clock period φ). The operands A and B are supplied via shift registers (SR) and the result, $S_3 \ldots S_0$, is stored step by step in a shift register at the output. We need only one full adder.

Bit-serial architectures need less hardware but they are slower than parallel architectures. Parallel data-path architectures are used for most modern VLSI processors (Berenbaum *et al.* 1987; Cushman 1988; Forsyth *et al.* 1987; Horowitz *et al.* (1987); Kadota *et al.* 1987; Kaneko et al. 1989; Miyake *et al.* 1990; Van Wijk *et al.* 1986). Bit-serial architectures are very suitable for pipelining in array processors and systolic arrays (Khung 1988). They can

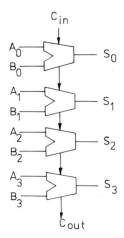

Fig. 4.11. Principle of a parallel adder.

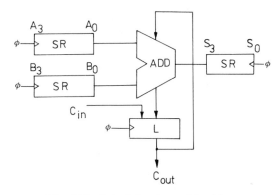

Fig. 4.12. Principle of a serial adder.

also run at higher clock rates and are very suitable for silicon compilation (Denyer and Renshaw 1985). A flowchart of a signal processing algorithm can be directly implemented in silicon with a network of bit-serial operators (functional blocks) (Denyer and Renshaw 1985). In this section we restrict our attention mainly to parallel data-path architectures. In Fig. 4.13 a general block diagram of a parallel data path is given. It has some functional blocks for input/output (latches and fast cache memories), register files, a shifter for bit-field manipulations, and an arithmetic-logic unit (ALU).

4.3.2 General-purpose processors

The data path of general-purpose processors must be very flexible to perform a wide range of different logical, arithmetic, and bit-field operations

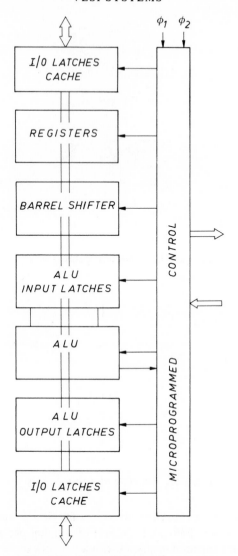

Fig. 4.13. General data path structure.

depending on the task running on the processor. In Fig. 4.14 we show a typical data path for an 8-bit microprocessor. It has only one data bus and, data and address computations are carried out in the same ALU (execution part) but the address bus is separated from the data bus by a tristate driver (couple). This offers the opportunity of incrementing the program counter PC and loading the memory address register while some other data processing task is taking place in the execution part.

In Fig. 4.15 a simplified data path of the 32-bit microprocessor μVAX

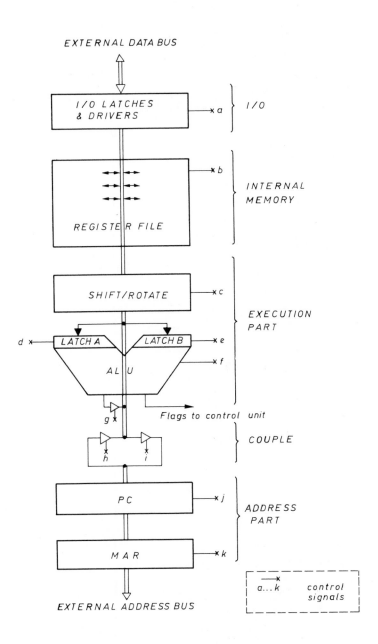

Fig. 4.14. An example of a simple data path.

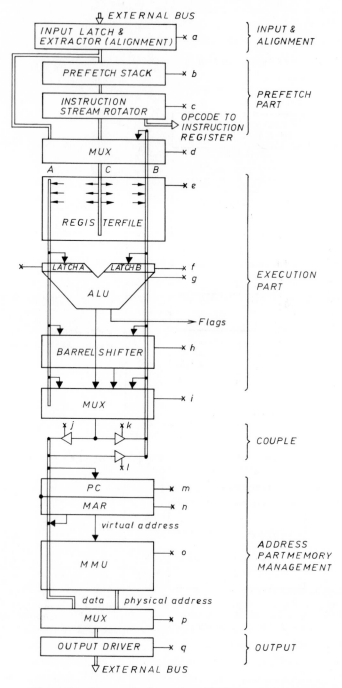

Fig. 4.15. Data path of a typical 32-bit processor chip.

32–780 (this is a registered trademark of Digital Equipment Corporation) (Supnik 1984) is sketched. Here we have three internal data buses A, B, and C, and one address bus. The data path can be subdivided into four main parts

(1) execution part;

(2) prefetch part;

(3) memory management part;

(4) input/output part.

Parallel processing is supported by the three data buses and the separation of the execution and the I/O parts. Pipelining is supported by a prefetch stack and an instruction stream rotator. The register file is a three-bus register file (see Fig. 3.44) and the ALU has the design shown in Fig. 3.37 with carry look-ahead techniques (see Fig. 3.36). The logical (virtual) address is translated into a physical (real) address in the memory management unit (see Section 4.5.6). The micro-operation signals controlling this data are labelled $a, b, c, \ldots, q$.

In Fig. 4.16 the principle of instruction prefetch using a cache is explained (Berenbaum *et al.* 1987; Forsyth *et al.* 1987; Horowitz *et al* 1987; Kadota *et al.* 1987). If an address is issued by the processor it is tested in the content – addressable part of the cache (CAM, see Sections 4.5.4 and 4.5.5) if the tags of the words (instructions) in the buffer part of the cache match the address issued by the processor. If a match is found then the instruction requested by the processor is already in the buffer part of the cache and can be loaded into the instruction queue. This saves the long time needed for an external memory access. (More details on cache will be given in Section 4.5.5.)

If no match is found a MISS signal is issued which leads to a replacement of a group (burst) of instructions in the cache.

The principle of pipelining will be explained with Fig. 4.17. Usually an instruction cycle consists of four steps:

(1) fetch (FE);

(2) decode (DEC);

(3) execute (EXE);

(4) data transfer (DAT).

Each individual step keeps only small portions of the whole data path busy so that more than one instruction can be processed at the same time. For example, execute (EXE) and fetch (FE) of compute instructions can be done at the same time. This is also true for data transfer and decode (see Fig. 4.17). However, for branch and load/store (memory) instructions it is somewhat more difficult. During execution of branch instructions a target address is

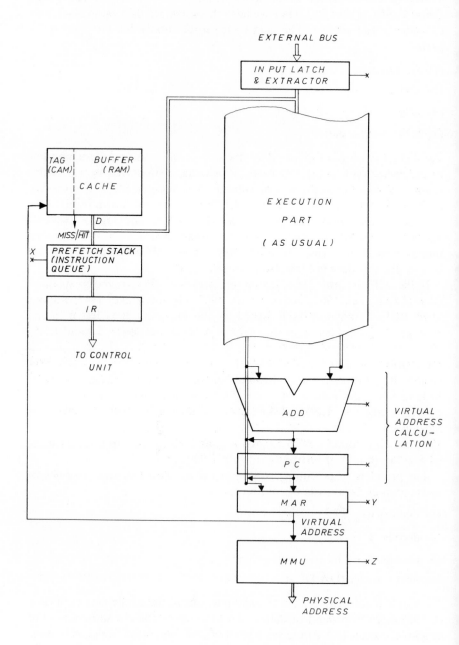

Fig. 4.16. Data path with prefetch cache.

PIPELINING

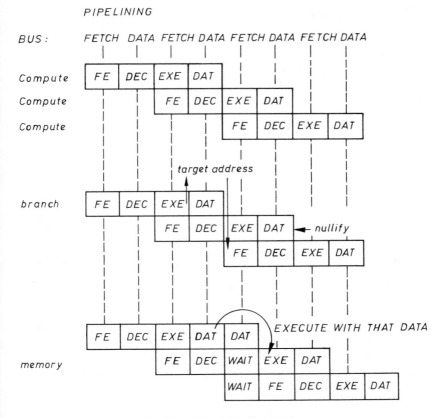

Fig. 4.17. Principle of pipelining.

computed but this is not usually the next address. Therefore the already prefetched next instruction is useless and must be nullified.

The same problem arises for memory instructions. The next (prefetched) instruction has to wait for execution (EXE) until the data transfer (DAT) of the current instruction is completed.

The data path of a RISC architecture has large register files because of their load–store principle. All compute instructions (see Fig. 4.9) have only registers as operand sources and destinations.

Some of the leading microprocessor families are (Crawford, 1986; MacGregor *et al.* 1984; Wilkinson 1986):

8-bit: Z80, 8080, 6800, 6502;

16-bit: 8086, 80286, 68000;

32-bit: 80386, 80486, 68020.

4.3.3 Digital signal processors (DSP)

The increasing demands of speed (real-time processing) and performance in modern digital signal and image processing cannot be fulfilled by general-purpose processors due to severe systems overload.

DSP algorithms are well-structured, regular, recursive and have local communication (Khung 1988). Typical computations for DSP are: for filtering

$$y_k = \sum_{n=0}^{N} a_n x_{k-n} + \sum_{n=1}^{K} b_n y_{k-n};$$ (4.1)

for image processing

$$y_{nm} = \sum_{j=0}^{N} \sum_{k=0}^{N} a_{jk} x_{n-j, m-k}.$$ (4.2)

The recursive sum-of-product computation is quite common in this type of DSP:

$$S_i = S_{i-1} + a_i x_i.$$ (4.3)

Therefore special data-path architectures and array processors have been developed (Van Wijk *et al.* 1986).

The data-path architectures considered in this section have precisely the same structure as those for general-purpose processors but without any overload. They are dedicated to executing the special recursive mathematical algorithms needed for digital signal or image processing. This means a high degree of parallel processing and hardware aids for complex mathematical operations, such as floating-point mutliplication and division. Some features of these data-path architectures are:

(1) a multibus system for parallel processing;

(2) address and data processing in different ALUs;

(3) functional blocks for complex mathematical operations;

(4) separate memories for data and instructions (HARVARD architecture).

In Figs. 4.18 and 4.19 two examples of data-path architectures for DSP are shown. The main features are:

(1) the operand sources for computations are very flexible because of the large number of multiplexers;

(2) the feedback loops via latches (previous data latch) are suitable for recursive algorithms;

(3) the multiplier and barrel shifter are hardware aids for the DSP algorithms.

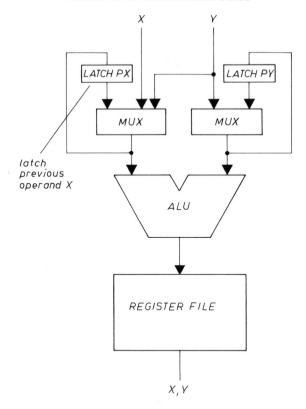

Fig. 4.18. Data path of digital signal processor, 1.

In Fig. 4.20 we show a data path for DSP with internal memory (ROM, RAM) for constants and variable data with a HARVARD architecture. Parallel processing is possible because of the multibus system.

4.3.4 Floating-point processors

Floating-point arithmetic done by general-purpose processors is very slow because of the many binary operations which are needed for a single floating-point computation. Therefore special data-path architectures have been developed (Benschneider 1989; Kaneko *et al*. 1989; Nakayama *et al*. 1989a; Takeda *et al*. 1985).

The data path for floating-point ADD and SUB in Fig. 4.21 is a hardware implementation of the algorithm (algorithm-based architecture). The floating-point operands X and Y are stored with sign (s), mantissa (m), and exponent (e) in the operand registers OPREGX and OPREGY. If the control signal ADD is 1 a floating-point addition is carried out. If ADD = 0

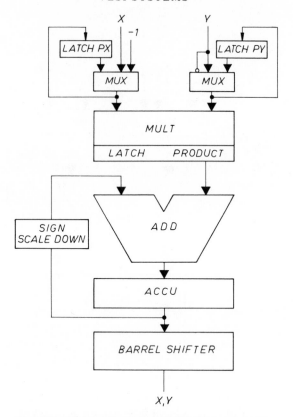

Fig. 4.19. Data path of digital signal processor, 2.

a subtraction is carried out by addition with the 2s complement of the mantissa of the subtractor X (depending on its sign s2). First the exponents e1 and e2 are compared in an adder:

$$B = e1 - e2. \qquad (4.4)$$

Then if $B < 0$ the carry CY is set ($CY = 1$), and if $B \geq 0$ the carry CY is reset ($CY = 0$).

We assume for our discussion that $e1 \geq e2$ ($CY = 0$). In that case e1 is transmitted via an incrementer INC into the result exponent register e. The mantissa m1 is transmitted to the adder ADD and the mantissa m2 is shifted B times right in a barrel shifter, because $B = e1 - e2 \geq 0$, and loaded into the adder ADD. Now the mantissa m1 is added to the shifted mantissa m2 and the result is normalized in a shifter. Depending on this shift the exponent must be incremented or left unchanged. Finally the sign, and occasionally the 2s complement, of the resulting mantissa must be computed as shown in Fig. 4.21 by the logic gates. For $ADD = 1$, we have

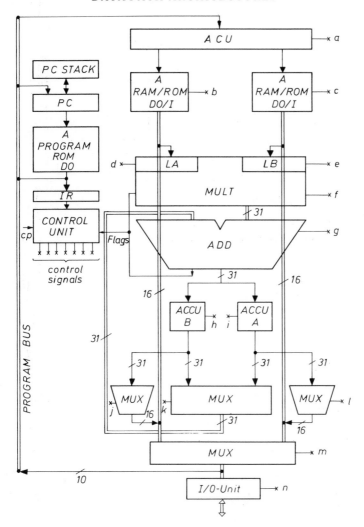

Fig. 4.20. Data path of digital signal processor, 3.

$s = 1$ if $s1 = 1$ and $s2 = 1$,
         or $s1 = 1$ and $s2 = 0$ and $CY' = 0$,
         or $s1 = 0$ and $s2 = 1$ and $CY' = 0$,

and $INC = \overline{s2}\ \overline{s1}\ CY' \vee s2\ s1\ \overline{CY'}$.

A floating-point multiplier can be designed in the same manner (see problem 7).

A quite different data-path architecture for floating-point multiplication and division is shown in Fig. 4.22. It is more like that of a general-purpose processor with special hardware resources for multiplication. The operands

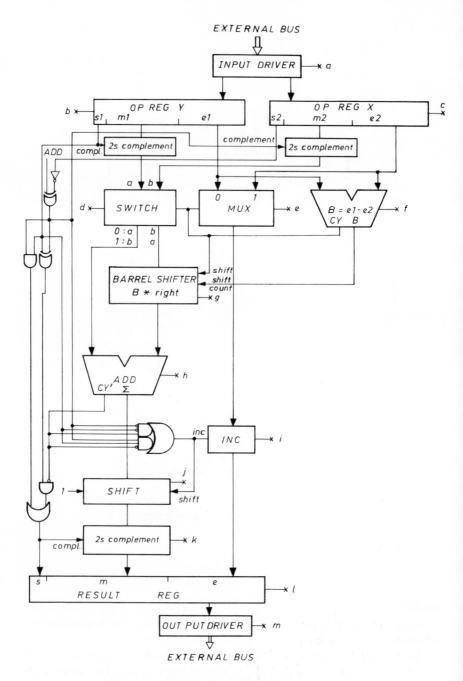

Fig. 4.21. Data path of floating-point processor, 1.

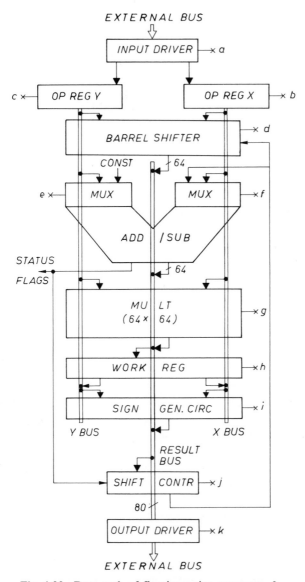

Fig. 4.22. Data path of floating-point processor, 2.

are loaded into the operand registers X and Y. The addition of the exponents is done in the full adder (ADD/SUB) simultaneously with the multiplication of the mantissas in the multiplier (MULT 64 × 64). This result may be temporarily stored in the work register. Finally a correction of the exponent bias and a normalization of the mantissa, together with a shift of the exponents in the barrel shifter, is carried out.

Table 4.3. Data flow in the floating-point processor data path of Fig. 4.22.

ADDITION

cycle	ADD/SUB	MULT	SHIFT CONTR	BARREL SHIFTER
1	comparison of the exponent			
2			number of shifts	
3				shift of mantissa
4	addition of the mantissas			
5	increment of exponent			normalization of mantissa

MULTIPLICATION

cycle	ADD/SUB	MULT	BARREL SHIFTER
1	addition of exponents	multiplication of mantissas	
2	correction of exponent bias		
3			normalization of mantissas

In principle this data path could also be employed for a floating-point addition or subtraction. In Table 4.3 we give a data flow model for floating-point ADD and MULT operations with the data path of Fig. 4.22. An overflow control is finally carried out.

4.3.5 Data processing arrays

Array processors are algorithm-based architectures which are likely to be important for future supercomputing. They are designed to reduce interconnection complexity and keep the architecture parallel and pipelined. In this section we will not discuss array processors generally; this is covered very extensively by Khung (1988). Instead we will simply indicate how semiconductor circuits can be used to implement data processing arrays.

These are one-, two-, or three-dimensional arrays of standard cells (processor elements) whose operation is determined by:

(1) the cell function;

(2) the location of the cell in the array;

(3) the interconnection structure;

(4) the data input and output.

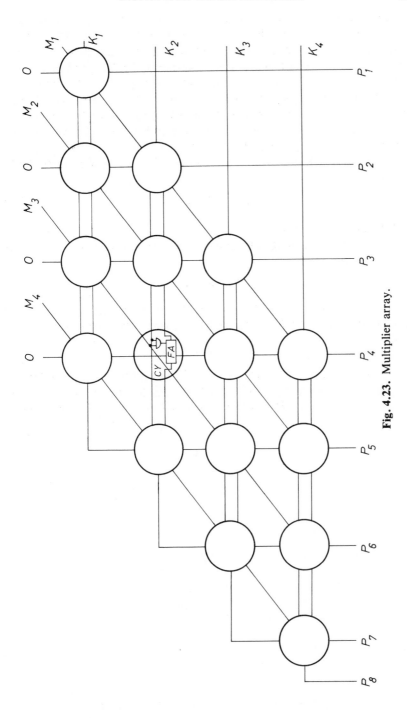

Fig. 4.23. Multiplier array.

Examples of these are matrix structures and systolic arrays. To clarify this we consider first, as a simple example, the multiplier array in Fig. 4.23. It implements the well-known pencil and paper multiplication algorithm (addition of shifted multiplier (Mano 1982; Santoro and Horowitz 1989)). The addition is implemented by the cell function (see Fig. 4.24), and the shift by the location of the cells in the array and the interconnection pattern (see Fig. 4.23). The cell in Fig. 4.24 is made with a full adder which adds an incoming carry C_j^i and two operands B_j^i and S_j^{i-1} to a sum S_j^i and creates a new carry C_{j+1}^i (see also Fig. 3.34). The bits of the multiplier K_j^i act as a control signal which determines the operand B_j^i given by the multiplicand M_j^{i-1}:

$$B_j^i = K_j^i M_j^{i-1} \tag{4.5}$$

(AND gate in Fig. 4.24).

For our simple 4×4 bit multiplier we need 16 cells and the result is the 8-bit product P

$$P = M \times K. \tag{4.6}$$

It works as follows. If the least significant bit of the multiplier, K_1, is 1 then the multiplicand M is added to 0 in the first row. The sum is now

$$S_4^1 S_3^1 S_2^1 S_1^1 = M_4 M_3 M_2 M_1. \tag{4.7}$$

If $K_1 = 0$ nothing will be added. Now in the second row an addition of the left-shifted multiplicand M and the current sum S^1 of row 1 is carried out if $K_2 = 1$. In our example $M = 1001$ (9) and $K = 0011$ (3) and we have

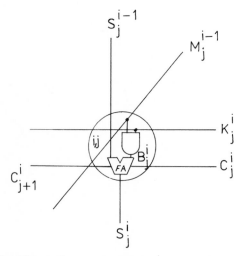

Fig. 4.24. Cell for the multiplier array of Fig. 4.23.

row 1. $K = 1$ 1001×1
         S 1001
row 2. $K = 1$ 1001×1
         S 11011
row 3. $K = 0$ 1001×0
         S 011011
row 4. $K = 0$ 1001×0
         S 00011011 (27)

The signal ripples through the array and in large arrays this can take a long time. Therefore carry look-ahead techniques should be applied.

Such hardware algorithms are faster than a software algorithm running on a general-purpose processor. Hardware implementation of algorithms saves computing time but needs a large area of silicon.

Systolic arrays are clocked (synchronous) systems. They need at least one host processor and the data flow is clocked through registers in between the cells. We can explain it by using a simple example (Fig. 4.25). For a sampled-data filter we have an input stream of sampled data $x_1, \ldots, x_n$. The filter characteristic may be given by $a_1, a_2, a_3, \ldots$ and we assume the filtered output data stream is computed by

$$y_i = \sum_{j=0}^{m} a_j x_{i-j}. \tag{4.8}$$

$$C_{j+1} = C_j + A_j B_j$$

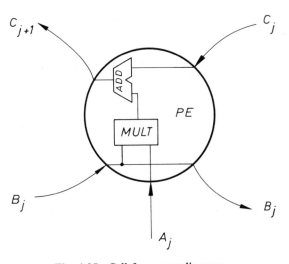

Fig. 4.25. Cell for a systolic array.

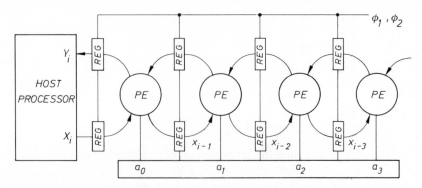

Fig. 4.26. An example of a systolic array.

For the hardware implementation of Equation (4.8) with systolic arrays we use processor elements (PE) which compute the inner product

$$C_{j+1} = C_j + A_j B_j. \tag{4.9}$$

The linear systolic array is shown in Fig. 4.26 in which the interconnection of the processor elements is done via clocked registers. The filter characteristic $a_1, a_2, \ldots$ is submitted by a data entry (DATA). Finally, the input and output sampled-data stream is received from and transmitted to a host processor.

4.4 THE CONTROL PATH

4.4.1 Principles

The task of the control path is to deliver micro-operation signals to the functional blocks of the data path (see Figs 4.14–4.16 *a, b, c, d, e, ...*) (Evanczuk 1984). All data-path operations can be understood as register transfer operations which can take place if a certain condition or control signal is true. In Fig. 4.27 we show this for a single register transfer operation. The control signal S (micro-operation signal) is a function of external signals $(X1, X2)$ and the inner state $(Z1, Z2)$. In our example we have

$$S = X_1 Z_1 \vee X_2 Z_2. \tag{4.10}$$

If $S = 1$ the register transfer operation (MOV A, B) is carried out.

In general we have, for the vector of micro-operation signals $\mathbf{S} = S1, S2, \ldots,$

$$\mathbf{S} = f(X, Z). \tag{4.11}$$

This is the operation of a finite-state machine called a Mealy automaton. If

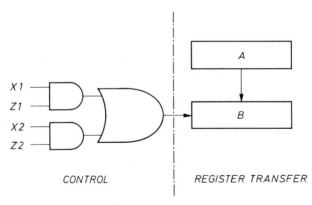

Fig. 4.27. Principle of register transfer.

the micro-operation signals **S** depend only upon the internal states **Z** we have a Moore automaton.

The operation of these automata is described with state or transition tables and state diagrams (see Fig. 4.28). The example of Fig. 4.28 describes the operation of a Mealy automaton with three internal states $\mathbf{Z} = Z1\,Z2 = 00$, 01, 11. The states are depicted in the state diagram with circles. The states are identified inside the circles. The transitions between the states are indicated with direct lines labelled with a vector of the input signals $\mathbf{X} = X1$ $X2, \ldots$. For example, the automaton performs a transition from state (11) to state (00) if the input signal vector is (00). The state transition table of our example of Fig. 4.28 is given in Table 4.4.

There are several styles for the hardware implementation of a control unit (Obreska 1982).

1. A random-logic control part is obtained by direct translation of the control logic equations $\mathbf{S} = f(\mathbf{X}, \mathbf{Z})$ into a set of logic gates and shift registers

Table 4.4. State table (x means 0 or 1).

input signal X		current state Z		next state Z	
x	0	0	0	0	0
x	1	0	0	1	1
0	1	1	1	1	1
0	0	1	1	0	0
1	x	1	1	0	1
1	0	0	1	0	1
0	0	0	1	0	0
x	1	0	1	1	1

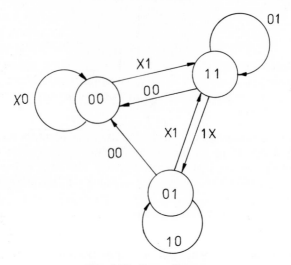

Fig. 4.28. State diagram.

RANDOM LOGIC

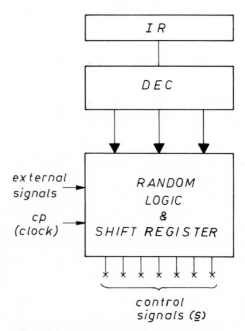

Fig. 4.29. Control unit with random logic.

(Fig. 4.29). To each state of the automaton there corresponds a stage in the shift register. The instruction decoder extracts the instruction properties which are used to validate the transitions in the shift register and parametrize the logical operations which yield the micro-operation signals (control signals). This method leads to a less regular control part which is hard to modify (redesign) during the design process but it gives the best silicon usage and the fastest control.

2. A slightly modified style is based on timing generators and random logic (see Fig. 4.30). Here an instruction is described as a set of micro-operations and a set of moments to activate them. The set of moments (clock states) are received from a timing generator which is controlled by property lines issued from an instruction decoder. The timer itself may be an automaton which gives the skeleton of the instruction execution.

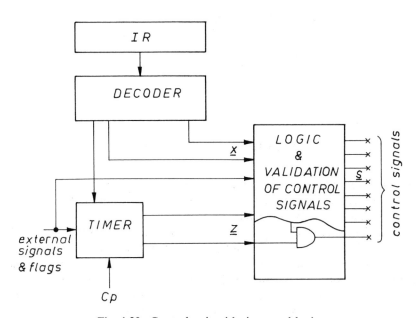

Fig. 4.30. Control unit with timer and logic.

3. The implementation of a Mealy automaton with PLA is a very regular one and is shown in Fig. 4.31. The input signal vector, **X**, is composed of external control signals and the decoded instruction signals. These input signals are supplied to the AND plane of the PLA. The signal lines Z1, Z2, ... representing the current state Z are also supplied to the AND plane. The micro-operation control signals **S** and the signals of the next state are issued by the OR plane. The programming of the PLA can be done with a

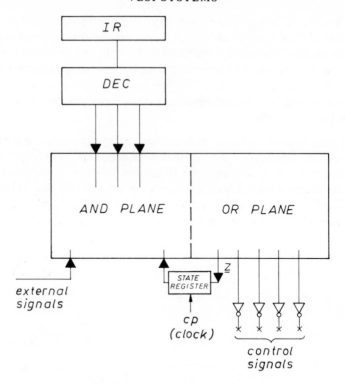

Fig. 4.31. Control unit with P L A.

computer and changes can be made very easily and in a short time. For large
PLAs we have large signal delays on the product-term lines so those control
parts with a single PLA are very slow.

   Multi-PLA control is more effective for complex control parts (Fig. 4.32).
Several PLAs for properties and parameter extraction, sequencing, and
micro-operation control − signal generation are applied. The property lines
are brought out directly from the instruction code and describe the global
characteristics of the instruction. The parameters describe local characteristics of a functional block as functions of the executed instruction. They are
extracted from the opcode and validated by a sequencing signal issued by the
command PLA which takes into account the current state and the properties
of the instruction code. The sequencing PLA computers the code of the next
state in the command PLA.

   4. Microprogrammed control
This method is quite different from those described so far. The main part
is a microprogram ROM (CROM, see Fig. 4.33) in which read-only
memory micro-instructions are stored. A micro-instruction has three parts:

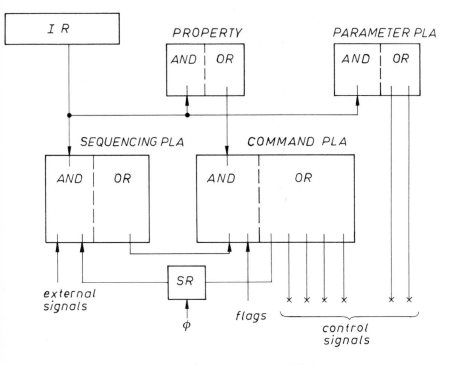

Fig. 4.32. Control unit with multi-PLA.

(a) control signal code;

(b) next micro-address;

(c) condition and sequencing control bits.

During a macro-instruction execution (e.g. ADD R1, R2) the microprogram sequencer steps the CROM through a sequence of micro-instructions which perform the macro-instruction. If the micro-instruction control codes issued by the CROM are the encoded control micro-operation signals for the functional blocks we call it horizontal microprogramming. If the micro-instruction control codes issued by the CROM are coded control signals which must be decoded by logic (PLA, see Fig. 4.34) we call it vertical microprogramming.

In CISC architectures, decoding is often performed by small local automata instead of local PLAs. The design of microprogrammed control means that the control store contents (microcode) must be defined and the sequencer logic, local decoders, and local automatons must all be specified. Because of its great importance we will now discuss this in greater detail.

HORIZONTAL MICROPROGRAMMING

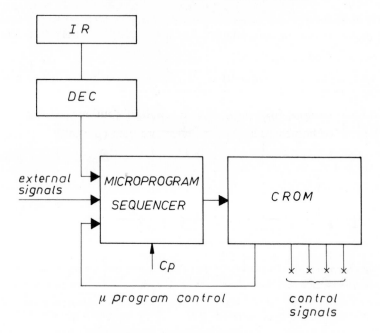

Fig. 4.33. Microprogrammed control unit.

VERTICAL MICROPROGRAMMING

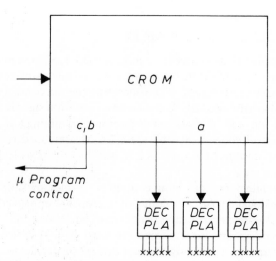

Fig. 4.34. Microprogrammed control unit (vertical programming).

4.4.2 Microprogrammed control

In Fig. 4.35 we give a detailed block diagram of a microprogram control part. It shows four options to generate an address for a micro-instruction:

(1) the next address (μNEXT);

(2) a start address for a sequence of micro-instructions performing a macro-instruction (μMAP);

(3) a return address from a micro-instruction subroutine (μSUB);

(4) a jump or branch address in the microprogram (μJMP).

One of these is supplied via a 1-of-4 multiplexer to the control address register (CAR).

The μNEXT address is generated by an increment of the current control address register content (CAR + 1) and stored in the microprogram instruction pointer (microprogram counter PC).

The μMAP address is decoded from the macro-instruction code by an address translator.

The μSUB address is popped from the microprogram address stack. At a microsubroutine call the current contents of the microprogram instruction pointer PC is pushed on to the stack. This will be the return address μSUB after the subroutine is completed.

The jump address c μJMP is contained in the micro-instruction word.

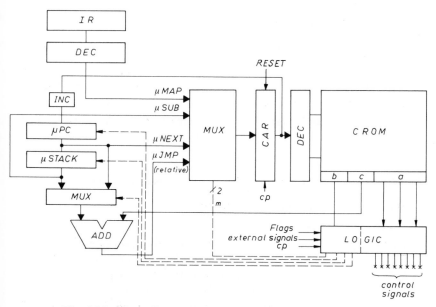

Fig. 4.35. Block diagram of a microprogrammed control unit.

This can be an offset for relative jumps (the offset is added to a base address) or an absolute jump address.

The selection of these four addresses is controlled by the sequencer logic shown in Fig. 4.35. The condition — select control signal bits b (contained in the micro-instruction word), flag bits (e.g. processor status bits for carry, sign, etc.), and external control signals (such as HALT, WAIT, INT) determine the two control signals $m = m1\ m2$ of the address multiplexer (e.g. $m1\ m2 = 00\ \mu MAP$; 01 μSUB; 10 $\mu NEXT$; and 11 μJMP). The control signals are decoded from the nanoword (a) of the micro-instruction.

4.4.3 Design example

System level (system definition)

We have to design a tiny 16-bit microprocessor which we call 'our tiny microprocessor' (OTM). The data format is 16-bit integer and the instruction format is 2 bits for the opcode and 14 bits for the operand. The instructions and the relating opcodes are listed in Table 4.5.

OTM uses a register-indirect addressing mode. The register file has two 16-bit registers R1 and R2 and the I/O signal lines are:

(1) 16-bit bidirectional data bus;

(2) 14-bit address bus;

(3) RESET signal line;

(4) HALT signal line;

(5) READ/WRITE signal line R/W.

Algorithmic level

The functional algorithm is described by the flowchart in Fig. 4.36. After RESET the processor halts in a no-operation (NOP) loop (state T_0). If no HALT signal is active the instruction fetch begins. During state T_1 the contents of the program counter PC are loaded into the memory address register MAR. During state T_2 the memory word (16 bits) at the location specified

Table 4.5. Instructions of our tiny microprocessor (OTM).

mnemonic	opcode	operation
NOP	00	no operation
ADD R1, (R2)	01	add the word in memory located at the address given in R2 to the word in R1
JPC (R2)	10	jump if carry is set to an address given in R2
STO (R2), R1	11	store the word in R1 to memory location given in R2

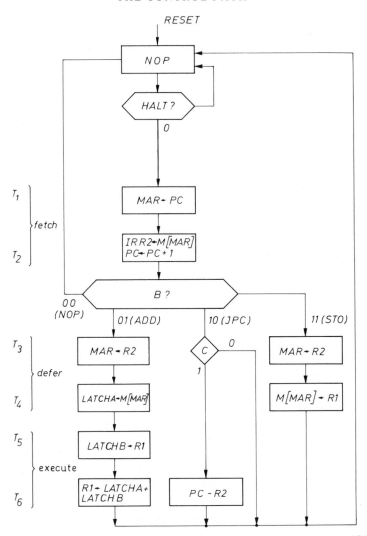

Fig. 4.36. Flowchart of the functional algorithm of our tiny microprocessor (OTM).

by MAR is read. The upper 2 bits (opcode) are loaded into the instruction register IR and the lower 14 bits are stored in register R2 (operand address). At the same instant the program counter is incremented ($PC = PC + 1$).

Now a decision is made depending on the opcode $B_2 B_1$. If $B_2 B_1 = 00$ (NOP) nothing is to be done and a jump to the start is executed.

If we have $B_2 B_1 = 01$ (ADD) an addition must be carried out. Therefore in the states T_3 and T_4 the operand is fetched from the memory location specified by the address given in R2 and is stored in LATCHA of the ALU. Next, in state T_5 the contents of register R1 are loaded into LATCHB of

Table 4.6. State codes of OTM.

T_0	0	0	0
T_1	0	0	1
T_2	0	1	0
T_3	0	1	1
T_4	1	0	0
T_5	1	0	1
T_6	1	1	0

the ALU, and finally, in state T_6, the addition is carried out in the ALU and the result is stored in register R1.

If the opcode is 10 (JPC) the carry is tested. If it is set the contents of register R2 (which is now the jump address) are loaded into the program counter PC. If the carry is not set (C = 0) then nothing is to be done.

If the opcode is 11 (STO) in state T_3 the contents of register R2 (which is a memory address) are loaded into the memory address register and in state T_4 the contents of register R1 are stored at the memory location specified by MAR (which was the contents of register R2).

We have used for this algorithm seven states $T_0 \ldots T_6$ which may be coded as shown in Table 4.6.

The flowchart boxes (states) can be directly mapped into microinstructions (microwords, see the discussion of the logic level below).

Function/register level

From the algorithm in Fig. 4.36 we obtain a final data-path structure with functional blocks, registers, and buses as shown in Fig. 4.37. Associated with these functional blocks is a set of micro-operation control signals $a_0, \ldots, a_{15}$. This data path is like that of Fig. 4.14. The register file has only two registers and we have only one flag, the carry flag C. The meaning of the micro-operation control signals $a_0, \ldots, a_{15}$ is explained in Table 4.7. The control path is designed as a microprogrammed control part.

Logic-level

Logic design means, in our example, design of the sequencer logic and definition of the microcode. The block diagram of our microprogrammed control part is shown in Fig. 4.38. It is a subset of the more general structure of Fig. 4.35. We have three types of microprogram addresses:

(1) μNEXT;

(2) μMAP;

(3) μJMP.

The micro-instruction word is 28 bits long and has a 16-bit nanocode $a_0 \ldots$

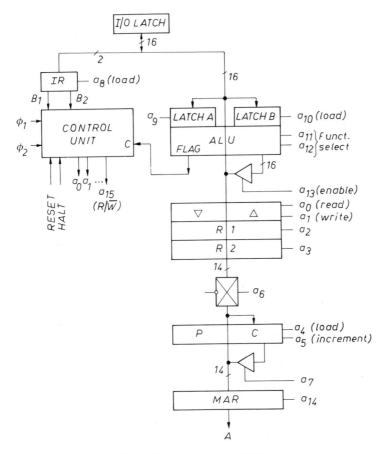

Fig. 4.37. Data path of OTM.

a_{15}, 4 condition select bits $b_0 \ldots b_3$ and 8 jump address bits $c_0 \ldots c_7$. The two flags are $F1 = \bar{C}$ (not carry) and $F2 = HALT$.

The truth table of the sequencer logic is given in Table 4.8. This can be implemented with random logic or with a PLA. The PLA definition, according to the methods treated in Chapter 3, is, for this sequencer logic

AND plane

$$
\begin{array}{cccccc}
- & 1 & 1 & 1 & - & - \\
0 & 0 & 0 & 0 & - & - \\
1 & 1 & 1 & 0 & 0 & - \\
1 & 1 & 0 & 1 & - & 0 \\
1 & 0 & 0 & 0 & - & -
\end{array}
$$

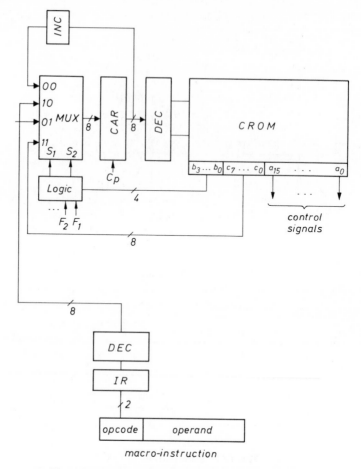

Fig. 4.38. Microprogrammed control unit of OTM.

OR plane

$$
\begin{array}{cc}
0 & 0 \\
1 & 0 \\
1 & 1 \\
1 & 1 \\
1 & 1
\end{array}
$$

(the reader should verify this).

The logic operation of the address translator is given in Table 4.9. The microcode can be generated by a microcode assembler but here we will do it 'by hand' for the macro-instruction JPC (R2) as an example. This code is located at address \$08 in the CROM (see Table 4.9). The first micro-insruction word at address \$08 is

Table 4.7. Micro-operation control signals of the OTM data path.

a_0	READ	read register
a_1	WRITE	write register
a_2	SELECT 1	select register R1
a_3	SELECT 2	select register R2
a_4	LOAD	load program counter
a_5	INC	increment program counter
a_6	COUPLE	couple data and address bus
a_7	PC ENABLE	enable program counter
a_8	LOAD	load instruction register
a_9	LATCHA	load LATCHA
a_{10}	LATCHB	load LATCHB
a_{11}, a_{12}	FSELECT1	function select ALU (11 for ADD)
a_{13}	ALU ENABLE	enable ALU to data bus
a_{14}	LOAD	load memory address register
a_{15}	R/$\overline{\text{W}}$	read–write control

Table 4.8. Truth table of the sequencer logic of the OTM microprogrammed control part.

b_3	b_2	b_1	b_0	F1	F2	S1	S2	
x	0	0	0	x	x	0	0	μNEXT
1	1	1	1	x	x	1	0	μMAP
0	0	0	1	1	x	1	1	μJMP
0	0	1	0	x	1	1	1	μJMP
0	1	1	1	x	x	1	1	μJMP

Table 4.9. Address translation.

OPCODE	CAR
00	$04
01	$0A
10	$08
11	$0E

    0001   00000000   0000000000000000
or, in hexadecimal notation,
    $b = \$1 \quad c = \$00 \quad a = \$0000$.
$b = \$1$ means jump if carry clear and the jump address is $c = \$0000$ while nothing is done in the data path $a = \$0000$ (see Fig. 4.37).
    The next micro-instruction is
    0111   00000000   0000000001011001

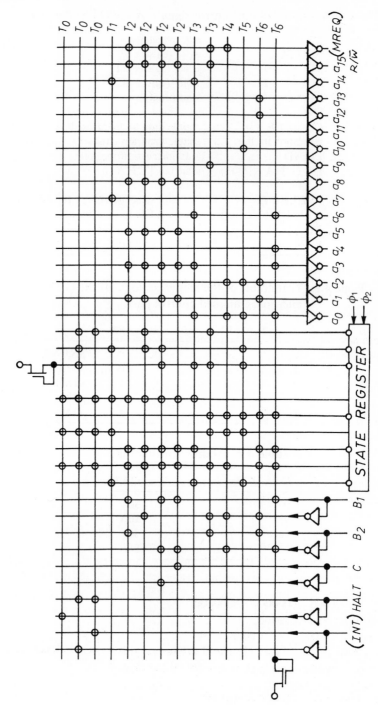

Fig. 4.39. PLA control unit of OTM.

or, in hexadecimal notation,

$b = \$7 \quad c = \$00 \quad a = \$0059.$

$a_0 = 1$, $a_3 = 1$ means read register R2, and $a_4 = 1$ and $a_6 = 1$ means load the data on the bus into the program counter. This is what we want to do (R2 → PC) if the carry is set (JPC (R2)). $b = \$7$ means jump unconditionally (see Table 4.8) and the jump address is $c = \$00$ (this means all from the beginning, see Fig. 4.36).

In Fig. 4.39 we show, for the same example, a PLA control unit corresponding to the design in Fig. 4.31. Here we have used two additional external signals, INT (interrupt) and MREQ (memory request). In problem 9 we ask the reader to verify this design.

Completion

Finally, in the design hierarchy of Fig. 4.3, the circuit and layout design follows from the four levels above.

4.5 SEMICONDUCTOR MEMORIES

4.5.1 Overview

There are several types of semiconductor memory. In Fig. 4.40 we give a generalized block diagram of a semiconductor memory. The memory is addressed by address lines for the row address and for the column address. The data transfer is carried on the bidirectional data lines $D_1 \ldots D_k$. The memory cells are arranged in a matrix (or several matrices) with M rows and N columns. With m row address lines we can have $2^m = M$ decoded row lines and with n column address lines we have $2^n = N$ column lines.

The two most important organizations are bit—organization where only one input and output line is available (DO/DI, see Fig. 4.40) and byte—organization in which 8 input/output lines are available at the periphery of the memory chip.

Depending on the physical behaviour and the structure of the memory cell we have five different types of memories

RAM (random-access memory) This memory can be randomly accessed. Read and write operations are possible with nearly the same access time. The memory cells are flip-flops for static RAM (SRAM) (Kobayashi *et al.* 1985; Komatsu *et al.* 1987; Matsui *et al.* 1987; Matsui *et al.* 1989; Miyaji *et al.* 1989; Sasaki *et al.* 1989; Tamba *et al.* 1989; Wada *et al.* 1987) and single capacitors for dynamic RAM (DRAM) (Chou *et al.* 1989; Fujii *et al.* 1989; Furuyama *et al.* 1986; Kimura *et al.* 1987; Kraus and Hoffman 1989; Mashiko *et al.* 1987; Taylor and Johnson 1985). In these memory cells the stored information (charge or potential) can be easily changed.

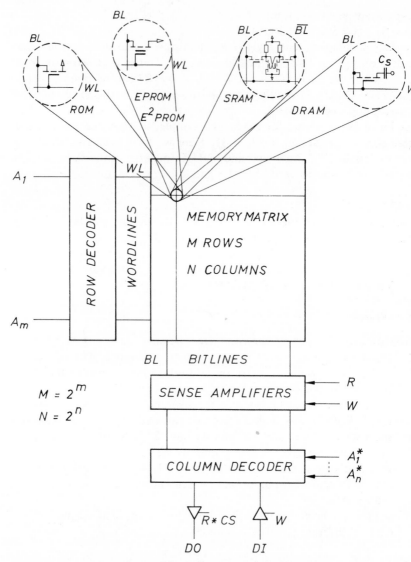

Fig. 4.40. Principle of semiconductor memories.

ROM (read-only memory) These memories can also be randomly accessed but the stored bit pattern cannot be altered so this memory can only be read. The memory cell is a single transistor which is fixed by the bit pattern (see Section 3.2.5); this cannot be erased.

EPROM, EEPROM (erasable programmable ROM) These memories can be randomly accessed and are usually read-only but they can be erased and electrically programmed. The memory cells are special MOS transistors with floating gates (Higuchi *et al.* 1990; Jolly *et al.* 1985; Kynett *et al.* 1988; Momodomi *et al.* 1989; Nakayama *et al.* 1989*b*; Samachisa *et al.* 1987; Ting *et al.* 1988).

CAM (content-addressable memory) These are associative memories where a location is indentified by comparison of the bit pattern stored in CAM with a given mask. This will be treated in detail in Section 4.5.4 (Goksel *et al.* 1989).

SAM (serial access memory) In SAM access to a specified bit or a memory word is only possible serially. CCD memories have such an organization. However semiconductor memories currently in use do not employ this method of access.

4.5.2 Random-access memories (RAM)

Static RAM

The memory cell of SRAM is a flip-flop as shown in Fig. 4.41 (Kobayashi *et al.* 1985; Komatsu *et al.* 1987; Matsui *et al.* 1987; Matsui *et al.* 1989; Miyaji *et al.* 1989; Sasaki *et al.* 1989; Tamba *et al.* 1989; Wada *et al.* 1987). The extremely high load resistors in Fig. 4.41(a) ($R > 500$ MΩ) are made with undoped polysilicon and are placed above the transistors. This gives a small cell area. The CMOS flip-flop of Fig. 4.41(b) has the advantage of small power and high α-particle immunity.

The memory cell is selected with a word line WL. If the potential of the word line is high the pass transistors are switched on and connect the flip-flop to the bit lines BL and $\overline{\text{BL}}$. In Fig. 4.42 we have sketched an SRAM cell corresponding to Fig. 4.41(a). These cells are arranged in $M \times N$ matrices on the chip. The bit lines run perpendicular to the word lines. The block diagram of an SRAM is shown in Fig. 4.43. Some of the address lines are fed into the row decoder (ROWDEC), and the others are fed into the column decoder (COLDEC). If $\overline{\text{CS}} = 0$ and $\overline{\text{WE}} = 0$ (where $\overline{\text{CS}} =$ chip select, $\overline{\text{WE}} =$ write enable) the data input lines DI are connected to the I/O amplifiers (COL I/O CIRC) whereas the data output lines are placed in their high-impedance state (tristate).

If $\overline{\text{CS}} = 0$ and $\overline{\text{WE}} = 1$ the data output lines DO are active and the bits of the selected memory word are available on the output lines.

If $\overline{\text{CS}} = 1$ DI and DO are placed in their high-impedance state the memory chip is disconnected from the system bus.

A network diagram of some important parts of the circuit of a static CMOS RAM is shown in Fig. 4.44 (Miyaji *et al.* 1989). The sense amplifiers are two-stage CMOS differential amplifiers as discussed in Section 2.2.6,

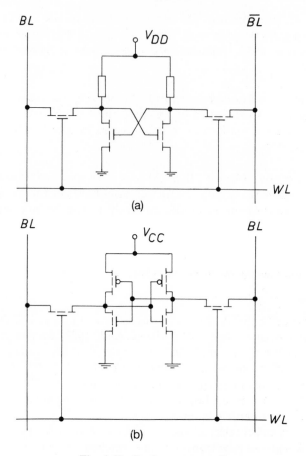

Fig. 4.41. Static memory cells.

Fig. 2.82. The bit lines are precharged at the beginning of each memory cycle. The selected memory cell (WL = 1, row select) is connected to the precharged bit line and causes a small potential difference ΔV between the bit lines BL and $\overline{\text{BL}}$. The selected columns are connected to two-stage CMOS differential amplifiers via transfer gates and ΔV is amplified.

The amplified read signal is transmitted via a tristate output driver (see Fig. 3.56) to the output PIN $D0_i$. The row- and column-select signals are generated by two-stage decoders as shown in Fig. 4.45 for the decoding of the address signals $A1, \ldots, Am$ to the row-select signals $WL1, \ldots, WLM$.

The timing diagram of the SRAM is depicted in Fig. 4.46. If the address is valid and the chip is selected ($\overline{\text{CS}} = 0$) the data on the output lines will be stable after an access time t_{acc}. If $\overline{\text{WE}} = 0$ a write cycle will be initiated. The write data must be stable on the data bus for a set-up time, t_s, before the leading edge of $\overline{\text{WE}}$ and for a hold time t_H after the trailing edge of the

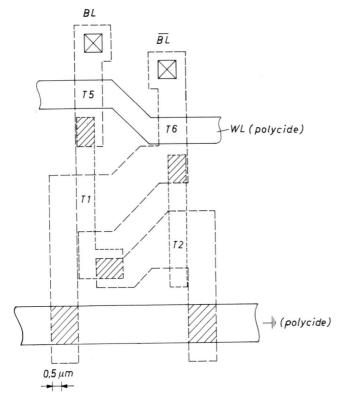

Fig. 4.42. Layout of a static memory cell.

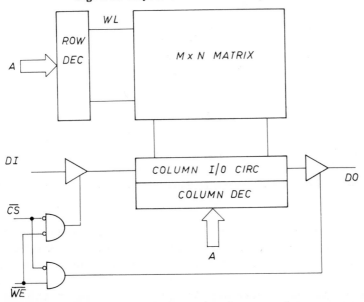

Fig. 4.43. Block diagram of a static semiconductor random access memory (SRAM).

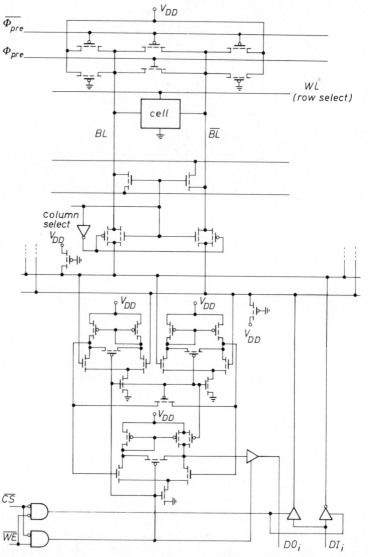

Fig. 4.44. Circuit diagram of the sense amplifier part of a static CMOS memory.

$\overline{WE}$ signal. The memory address must be valid during the memory cycle t_c.

Fig. 4.47 shows the PIN assignment of a 1 Mbit SRAM. It has 17 address lines A0, ..., A16 and 8 bidirectional data lines (byte organization 128k × 8).

BICMOS technology (see Section 3.1.1) yields very fast SRAMs (Tamba *et al.* 1989). The fastests SRAMs on silicon today are ECL or CML SRAM

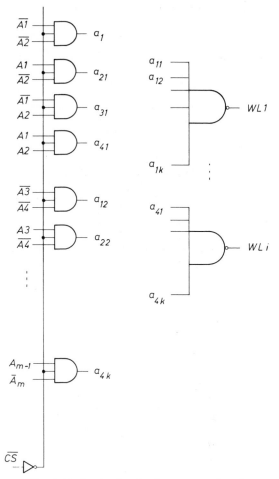

Fig. 4.45. Logic circuit of two-stage decoder.

(Matsui *et al.* 1989). Three-dimensional SRAM designs on laser recrystal-lized polysilicon layers give a high memory density (see Fig. 4.48, (Inoue *et al.* 1986)). The memory cells are located on the first floor and the decoder, buffer, I/O driver, and sense amplifier on the second floor as shown in Fig. 4.49.

Dynamic RAM
Modern dynamic memories employ, as a memory cell, a capacitor C_S selected via a pass transistor (single transistor cell), see Fig. 4.40, (Chou *et al.* 1989; Fujii *et al.* 1989; Furuyama *et al.* 1986; Kimura *et al.* 1987; Kraus and Hoffmann 1989; Mashiko *et al.* 1987; Taylor and Johnson 1985). For memory densities of up to 1 Mbit, a planar plate capacitor was applied

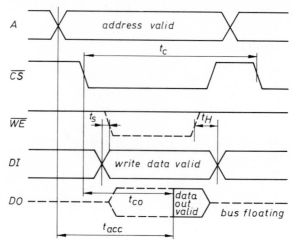

Fig. 4.46. Timing diagram of a static semiconductor memory (SRAM).

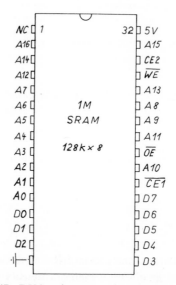

Fig. 4.47. PIN assignment of a 1 Mbit static RAM.

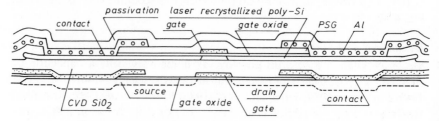

Fig. 4.48. Cross-section of a three-dimensional SRAM.

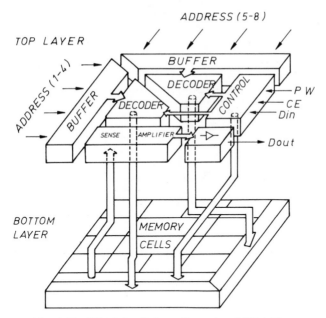

Fig. 4.49. Principle of three-dimensional SRAM.

(Watanabe *et al.* 1989). Because of leakage currents and the need for α particle immunity the stored charge on the capacitor C must be at least $Q = 200$ fC.

Taking into account a maximum field in the SiO_2 of $E_m = 2 \times 10^6$ Vcm^{-1}, well below the breakdown, the minimum area of the storage capacitor C is given by

$$A_c = \frac{Q_m}{\epsilon_H E_m} = \frac{2 \times 10^{-13}\,cm^2}{10^{-12} \times 2 \times 10^6} = 10\,\mu m^2 \qquad (4.13)$$

where $\epsilon_H = 10^{-12}$ As/Vcm is the permittivity of silicon. However for a 16 Mbit DRAM the whole cell size, including the pass transistor, must be no larger than 5 μm^2 (Fujii *et al.* 1989) so the third dimension must be used to satisfy the requirements of future DRAMs.

Several kinds of stacked and trench capacitor cells have been developed for multimegabit memories (Chou *et al.* 1989; Fujii *et al.* 1989; Kimura *et al.* 1987). And in Fig. 2.33 we have already considered a trench transistor cell. The layout and a cross-section of a simple trench capacitor cell is shown in Fig. 4.50. Here the capacitor plate is polysilicon which fills a trench groove. The gate of the pass transistor on the left side is molybdenum and is connected to the word line. The bit lines run in aluminium. The p-well impurity concentration is designed to suppress leakage currents and to give a good α-particle immunity. The bit line is precharged to $V_{DD}/2$ and has a

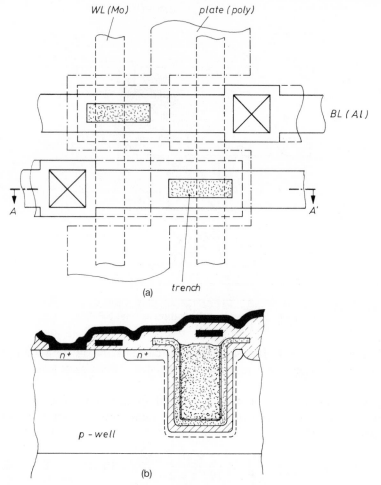

Fig. 4.50. (a) Layout and (b) cross-section of a trench cell for dynamic megabit memories.

capacitance C_B. If the memory cell is selected (connected to the bit line) a small potential change, ΔV_B, occurs on the bit line

$$\Delta V_B = \frac{C_s}{C_s + C_B}(V_C - V_B) \qquad (4.14)$$

where V_B is the bit-line potential and V_C is the cell potential. Here $C_s/C_B \ll 1$ and therefore ΔV_B is of the order of some 10 mV, and $(V_c - V_B) \approx V_{DD}/2$.

The sense amplifier circuit is shown in Fig. 4.51 together with some memory and dummy cells (Furuyama *et al.* 1986). The capacitor plates and

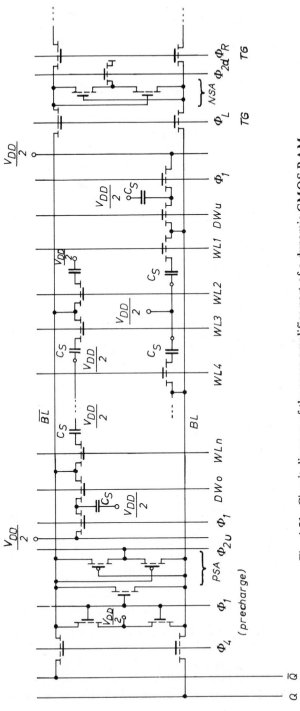

Fig. 4.51. Circuit diagram of the sense amplifier part of a dynamic CMOS RAM.

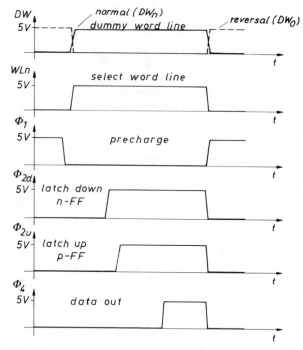

Fig. 4.52. Timing diagram of the read-out cycle of a dynamic RAM.

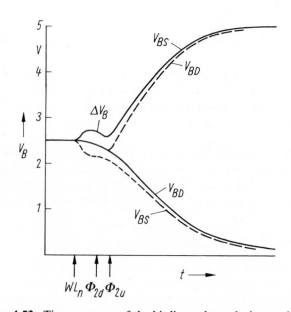

Fig. 4.53. Time response of the bit line voltage during read-out.

the folded bit lines are precharged to $V_{DD}/2$ to relax the electric field in SiO_2. The dummy cells are also precharged to $V_{DD}/2$. The sense amplifiers consist of NMOS (NSA) and PMOS (PSA) parts. With the internal timing diagram in Fig. 4.52 we can describe the operational principles as follows.

With clock Φ_i the bit lines and the dummy cells are precharged to $V_{DD}/2$. If we select (activate) a word line WLn we have a small potential increase or decrease, ΔV_B, (depending on whether a '0' or a '1' is stored in the cell) on the corresponding bit line. Now in the clock phase Φ_{2d} a latch-down via the NSA part is realized. A short time later the latch-up via the PSA part takes place (clock phase Φ_{2u}). This leads to a potential V_{DD} or 0 on the bit lines (see Fig. 4.53) depending on whether the initial potential change ΔV_B was positive or negative. During clock phase Φ_4 the data is read out. To reduce the capacitive load of the NSA the left- and right-hand sides of the bit line are coupled via pass transistors (controlled by Φ_R and Φ_L) to the NSA part. The floorplan of Fig. 4.54 shows the location of the NSA and PSA parts in a 4 Mbit DRAM.

Another version of a floorplan is shown in Fig. 4.55. The Y-decoders are placed in between the sense amplifiers and the X-decoders are in the middle of the chip.

The block diagram of a DRAM is shown in Fig. 4.56 where the memory cells are arranged in $M \times N$ matrices. The control signals are similar to those for SRAM. The memory address is split into two parts, a row address and a column address. The row address is supplied first and strobed with the row-address select signal $\overline{RAS}$ into the row-address buffer. Some time later ($t_{RCD} \approx 20$ ns) the column address is supplied and strobed with the column-address select signal $\overline{CAS}$ (see Fig. 4.57). The action is triggered on the falling edge of the $\overline{RAS}$ and $\overline{CAS}$ signals and they are kept low to the end of the cycle. The read cycle is initiated by $\overline{WE} = 1$ when $\overline{CAS}$ becomes low. The time before the data appears stable at the output line DO from the falling edge of $\overline{RAS}$ or $\overline{CAS}$ is called the row-address or column-address access time, t_{RAC} and t_{CAC} respectively. The data remains valid while $\overline{CAS}$ is low. If $\overline{CAS}$ goes high the data output lines are placed in their high-impedance state (tristate).

For a write operation $\overline{WE}$ is set to 0. If $\overline{WE}$ goes low before $\overline{CAS}$, as shown in Fig. 4.57, we have an early write operation and the data must be valid before the falling edge of $\overline{CAS}$. The data is strobed with $\overline{CAS}$.

The PIN assignment of a 4 Mbit DRAM is shown in Fig. 4.58.

4.5.3 ROM, EPROM, and EEPROM

Read-only memories have the same organization and control lines as RAM but they have quite different memory cells. A ROM cell is just a single transistor which is placed in a memory matrix and is connected or not connected to the word and bit lines depending on the bit pattern stored in the memory.

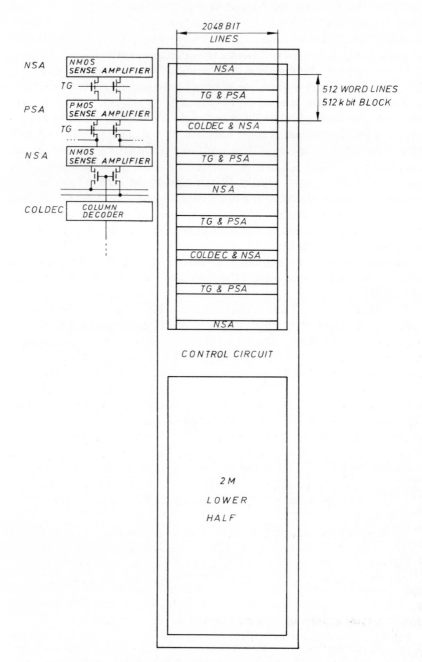

Fig. 4.54. Floor plan of a dynamic 4 Mbit memory chip.

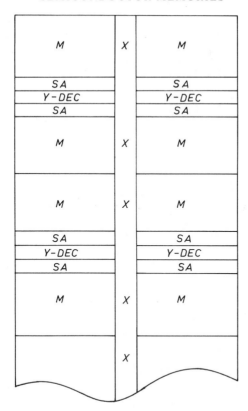

Fig. 4.55. Floor plan of a memory chip.

In Fig. 4.59 we show a layout and cross-section of a ROM memory cell.

EPROM are erasable programmable read-only memories. They have, as memory cell, a floating-gate transistor (SAMOST) (Higuchi *et al.* 1990). The floating gate, FG, can be charged with hot electrons from the semiconductor surface. If we supply a high voltage (20 V) at the bit and word line in the programming mode, a hot electron current flows across the Si–SiO$_2$ barrier on to the floating gate (see Fig. 2.64) and charges it negatively. Such a charged floating-gate transistor has a much higher threshold voltage ($V_{tH} \approx 10$ V) than an uncharged transistor ($V_{tL} \approx 1$ V). This gives the opportunity to read that device with a gate potential V_G which is between V_{tL} and V_{tH}. An uncharged transistor gives a current flow ('1'); the other does not ('0'). EPROM cells can be erased with ultraviolet light.

Flash EEPROM cells have a very thin ($\approx$ 20 nm) SiO$_2$ layer between the floating gate and the semiconductor surface. The control gate forms an enhancement transistor in series with the floating – gate transistor as shown

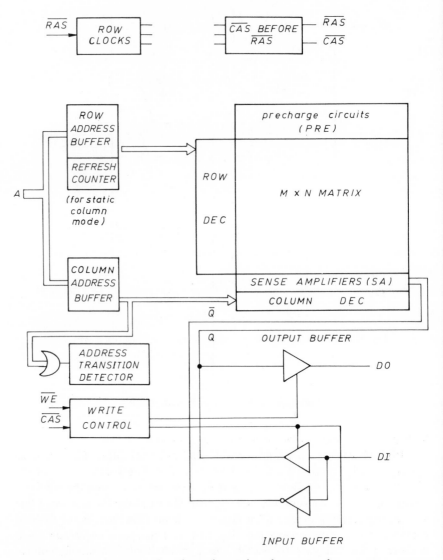

Fig. 4.56. Block diagram of a dynamic semiconductor random access memory (DRAM).

in Fig. 4.60 (Kynett *et al.* 1988; Samachisa *et al.* 1987). These flash EEPROM can be electrically erased. Other EEPROM cells consist of two separate transistors, a floating – gate transistor and a pass transistor for cell selection (Jolly *et al.* 1985; Ting *et al.* 1988). This is shown in Fig. 4.61.

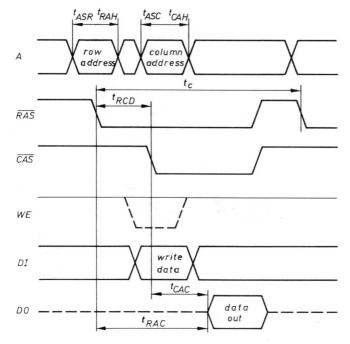

Fig. 4.57. Timing diagram of a dynamic semiconductor RAM.

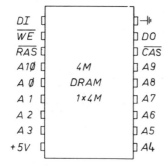

Fig. 4.58. PIN assignment of a 4 Mbit dynamic RAM.

4.5.4 Content-addressable memory (CAM)

In a content-addressable memory (or associative memory) a location is identified by its contents rather than by an address. The data word applied to the inputs, called a mask m, are compared simultaneously with all data words b stored in the CAM. If any word is found which is the same as the applied word a match signal e is generated. The logic diagram of a memory cell for

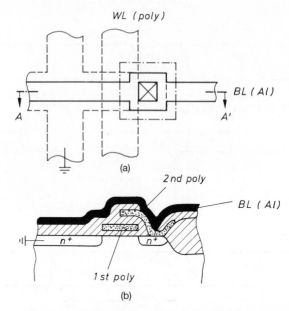

Fig. 4.59. (a) Layout and (b) cross-section of a ROM cell.

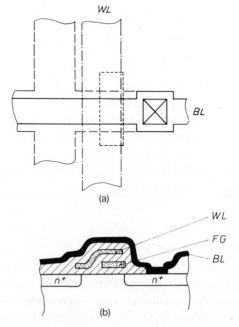

Fig. 4.60. (a) Layout and (b) cross-section of a flash EPROM cell.

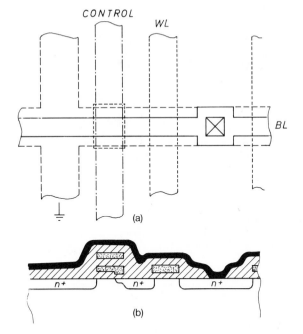

Fig. 4.61. (a) Layout and (b) cross-section of an EEPROM cell.

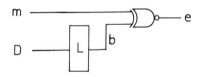

Fig. 4.62. Associative memory cell.

CAM is shown in Fig. 4.62. It has a store function represented by the latch L and a compare function done by the exclusive NOR gate.

A CMOS circuit of a CAM cell is shown in Fig. 4.63. The latch is a CMOS flip-flop and the bit lines, BL and $\overline{BL}$ are used for data replacement, D and $\overline{D}$, and the mask bit, m and $\overline{m}$.

4.5.5 Cache

A cache is an on-chip high-speed memory that alleviates the problem of speed mismatch between processor and external semiconductor memories (Miyake *et al*. 1990; Sawada *et al*. 1989; Wilkinson 1986). It is usually associatively mapped, in which case a content-addressable memory is coupled to a random-access memory (buffer) as shown in Fig. 4.64. To every word stored in the RAM part (buffer) there corresponds a label called TAG

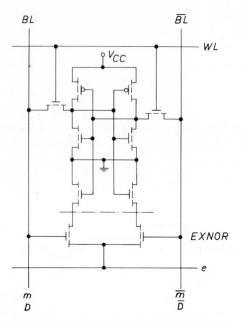

Fig. 4.63. CMOS circuit of an associative memory cell.

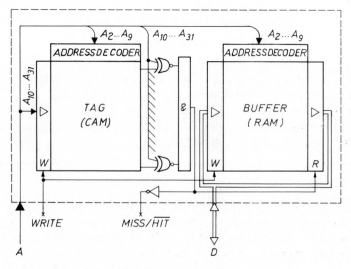

Fig. 4.64. Content-addressed memory (CAM).

in the CAM part. If a match is found in the CAM part the corresponding word in the RAM part is read out. If no match is found a MISS signal is generated informing the processor that the word which has been requested is not available in the cache. In that case the operating system has to run a replacement mechanism (swapping) as we will see a little later. In our example of Fig. 4.64 the labels (TAG) assigned to each data word stored in the buffer are the upper-address bits $A_{10}...A_{31}$ and the entries in the TAG register file are given by the lower bits $A_2...A_9$. The comparison is made by exclusive NOR gates and a NAND as we have shown in Fig. 4.62.

The data in the TAG and the buffer can have different meanings.

1. The TAG may be part of the physical address and the consecutive data or instruction words (or groups of these) are stored in the buffer. In this case a cache is successful because programs usually reference instructions near previous memory references and particular instruction and data references are often repeated. Therefore once a cache is loaded from the primary memory it is used more than once before data must be replaced (Sawada *et al.* 1989).

2. The TAG may be part of the virtual address and the consecutive physical addresses are stored in the buffer (RAM). In that case the cache is used as a translation look-aside buffer (TLB). This saves the long time needed for virtual-to-physical address translation (Goksel *et al.* 1989; Leilani *et al.* 1990).

The replacement mechanism of our cache is very poor; every word which is not found will be replaced. A more sophisticated replacement mechanism is possible with the cache shown in Fig. 4.65. It employs the so-called least recently used (LRU) replacement mechanism: the least recently used data in the cache will be replaced. The hardware in Figs 4.65 and 4.66 works as follows.

With the full associative mapping shown in Fig. 4.66, the match signal is now a match word e with m bits (in Fig. 4.64 it was just a single MISS/$\overline{\text{HIT}}$ signal line). This match word is used to generate a page fault signal. Page fault = 0 means there was a match; page fault = 1 means the requested data is not available in the cache. If a match is found the match address e is the uncoded mapping address into the buffer (via MUX1) and the data can be read out. Simultaneously this address is pushed via MUX2 on to the top of the LRU register (this will be explained with Figs 4.68 and 4.69 later).

If no match is found the external main memory must be referenced. The buffer is now addressed with the uncoded least recently used address a' coming out of the LRU register (see below) via MUX1 (because the page fault signal is now 1). The write signal is active and a new data entry can be made from the data bus into the buffer of the cache at exactly that location where formerly the least recently used data was stored. The address of the buffer

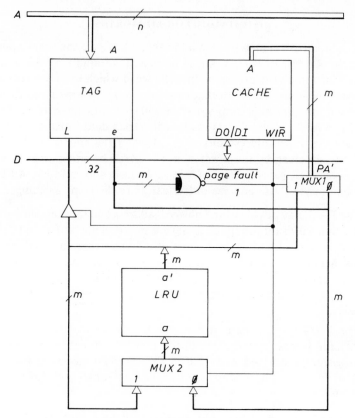

Fig. 4.65. Cache with LRU replacement.

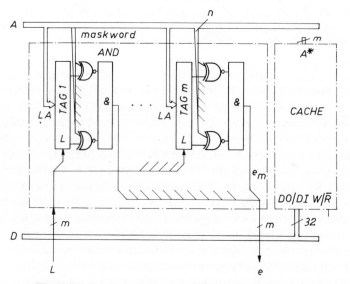

Fig. 4.66. TAG register part of the cache in Fig. 4.65.

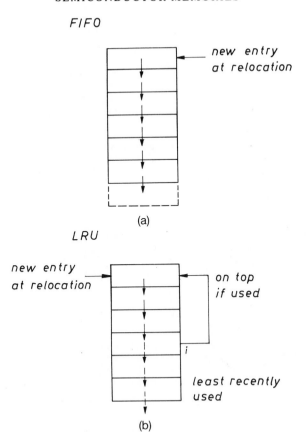

(a)

(b)

Fig. 4.67. Replacement mechanisms of a cache: (a) FIFO and (b) LRU.

which is now used (the least recently used) will be simultaneously pushed on to the top of the LRU register via MUX2 and becomes the most recently used. The TAG is also updated with the corresponding address on the address bus in the recently used TAG register. The uncoded address a' (as for the buffer) is the load signal for the recently used TAG (see Fig. 4.66).

We will now describe in greater detail the LRU principle (see Fig. 4.67). If a buffer address is used it is removed from its current location i in the LRU register and pushed on to the top (new entry) and all addresses in the LRU register above that location i are moved one position down in the LRU register. This is in contrast to the FIFO principle also shown in Fig. 4.67. Here the address which is in the queue for the longest time is transferred out. As a new entry is inserted all entries move down one place and the last is used for replacement. A block diagram and a logic circuit of the LRU register is shown in Figs 4.68 and 4.69, respectively.

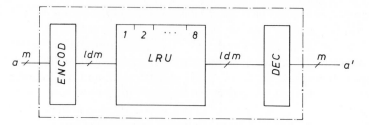

Fig. 4.68. Block diagram of the LRU unit.

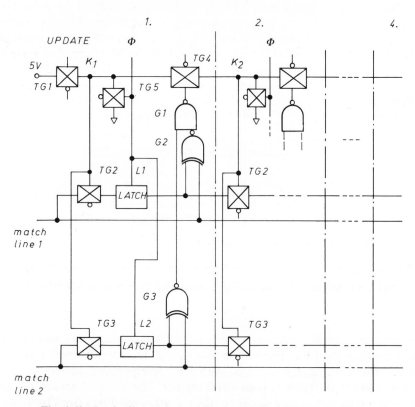

Fig. 4.69. Logic diagram of the LRU circuit. Φ is the latch clock.

The uncoded m bit buffer address a will be coded at the entry of the LRU register and decoded at the output a' of the LRU register. To explain this operation we use a simple example with $m = 4$ or two input match lines of the LRU register in Fig. 4.69. The latches L1 and L2 form the 2-bit LRU register cell, the signal line UPDATE activates an LRU mechanism and the node K_1 will be precharged. Now the match lines are compared with the contents of the latches in every LRU register stage with exclusive NOR gates

G3 and G2 shown in Fig. 4.69. If a match is found the output of the NAND gate G1 is 0 and the transfer gate TG4 is switched off.

Therefore all successive nodes $K_2, \ldots$ are kept low and no shift down of the successive LRU register contents via the transfer gates TG2 and TG3 can happen. If no match is found TG4 is switched on and the following node (here K_2) is charged up to 5 V. Now a shift down one place via TG2 and TG3 will be done. At the same time a new address is loaded into the first (top) LRU register L1 L2 from the match lines because node K_1 is high. We will clarify this further by considering a numerical example.

We assume in stage 1 (top) L1 L2 = 01, and in stage 2, L1 L2 = 11. The buffer of the cache should be referenced with the coded address 11 so we have, on the match line, the bit pattern 11. This does not match with 01 in the first stage of the LRU register. Therefore the 01 of the first stage is shifted via TG2 and TG3 into stage 2. But the contents 11 of stage 2 matches with 11 at the input. Therefore this is not shifted downward into stage 3. Simultaneously the 11 is loaded into the L1 L2 of the first stage.

In summary we have

stage	before	after	
1	01	11	top
2	11	01	
3	00	00	
4	10	10	least recently used

4.5.6 Memory management unit (MMU)

The task of the MMU is to translate a logical or virtual address (LA) into a physical address (PA), to report page faults and to give some protection

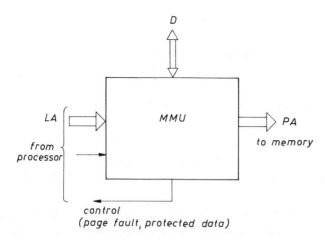

Fig. 4.70. Memory management unit.

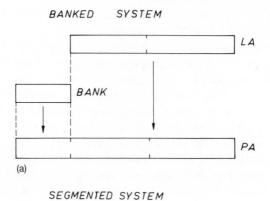

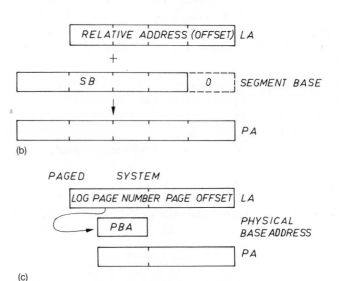

Fig. 4.71. Principles of address translation: (a) banking, (b) segmentation, and (c) paging.

(see Fig. 4.70) (Mano 1982). Some simple principles of address translation mechanisms are sketched in Fig. 4.71.

1. The banked system is used in 8-bit processors if they will reference more than $2^{16} = 64$ kbyte. The bank (e.g. 8 bits) may be an external I/O register and is concatenated with the 16-bit address bus of the processor to give a 24-bit address.

2. The segmented system divides the memory space into variable-sized blocks called segments. Each address is composed of a segment base and an offset. They are added together to give the physical address (PA).

3. In the paging system the memory space is divided into blocks of equal size (e.g. 512 bytes for the VAX; 4096 bytes for the 80386); each block is called a page. Each address is composed of a page number (physical base address) and a line number (page offset) which are concatenated to give the physical address PA.

In Fig. 4.72 we show an MMU which supports the paging system. The logical (virtual) address is composed of a page offset and a logical page number. The logical page number is an index into the page frame table. The page frame table is updated by the operating system and contains the physical base address PBA, protection bits (which validate the access rights), and a valid bit (which indicates that the page is currently in the primary semiconductor memory). The PBA is concatenated with the page offset to give the physical address PA.

The segmented system in Fig. 4.73 uses the upper bits of the logical address LA (segment number) as an index into a segment descriptor table. Every segment descriptor consists of a physical base address, protection bits, and a validity bit. The protection bits also contain information on segment length, segment limit, and segment usage. The physical base address can be added to an offset, as already shown in Fig. 4.71, or to a page number and concatenated with the page offset to give the physical address PA, as shown in Fig. 4.73. Every segment contains a variable length of 1 to 2^P pages and every logical page can be located at any location in the physical memory space.

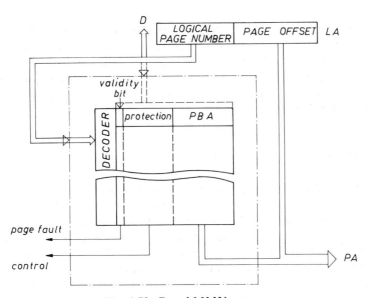

Fig. 4.72. Paged MMU system.

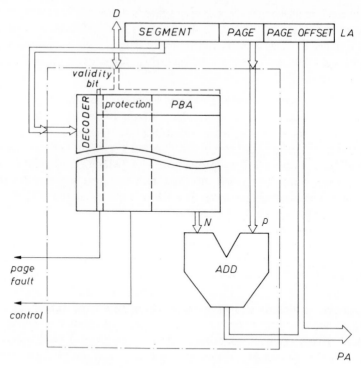

Fig. 4.73. Segmented MMU system.

Each logical-to-physical address translation needs at least one memory access (the 80386 needs three!). Therefore cache memories are employed as translation look-aside buffers (see Section 4.5.5) (Crawford 1986). If a page is not in the semiconductor RAM, the MMU issues a page fault (MISS) and the operating system runs an exception handling routine called swapping, illustrated in Fig. 4.74.

4.6 PROBLEMS OF TESTING

Testing is of growing importance for complex integrated semiconductor systems and is significantly influencing future designs (Williams 1986). In Fig. 4.75 we give an overview of the problems of testing. The stimuli used for testing are test-patterns. These can be exhaustive (all possible) or a limited number generated randomly or by fault modelling. The output pattern of the real machine is compared with the output pattern of the good (simulated) machine.

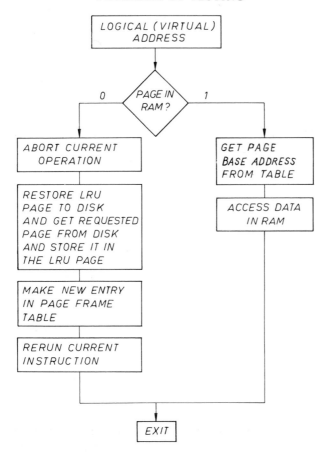

Fig. 4.74. Algorithm of memory swapping.

4.6.1 Fault models

Because of the enormous number of test-patterns needed for exhaustive testing to ensure that a complex system is free of defects or failures it would take years to complete a functional test of a complex system. But if we had some information about the structure of the implementation of the system we could apply a relatively small number of tests to ensure that a given set of faults did not exist. Fault models can help to solve this problem.

On a semiconductor chip we can expect to find a large number of faults and one of the most widely used fault models is the stuck-at model. In this model it is assumed that physical defects and faults will result in a permanent value of 0 or 1 on the logic lines (S-A-0, S-A-1), irrespective of what value

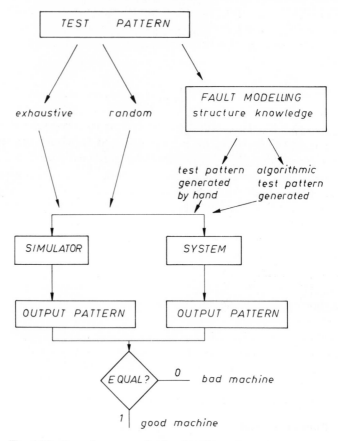

Fig. 4.75. Overview on methods of testing using a test pattern.

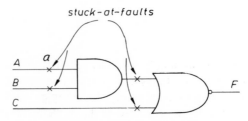

Fig. 4.76. Stuck-at-fault model.

is applied. In the fault simulation it is assumed that only one fault occurs at a time. To demonstrate this let us look to the simple example in Fig. 4.76. This circuit has three inputs so for exhaustive testing it needs $2^3 = 8$ test-patterns. If we assume an S-A-0 at node (a) only 2 test-patterns, $ABC = 010; 110$, are needed to see whether the fault exists.

Every reduced test-pattern (by fault modelling or by random generation) leads to an incomplete fault coverage so, for VLSI systems, design-based test methods are preferred. These are design for testability and built-in self-test.

4.6.2 Design for testability

To these methods belong all *ad hoc* techniques where special test facilities are implemented on the chip. More general are the level-sensitive scan design methods (LSSD) where a scan path is designed to reduce the sequential depth of the system. The principle will be explained with Fig. 4.77. The memory elements are separated from the combinational logic which is fed by primary inputs and the memory elements. The memory elements are observed via a shift register (scan out). Test-patterns for the combinational logic can come from a test generation program that generates tests only for the combinational part rather than the more difficult case of sequential logic (Williams 1986).

If the memory elements filed in an array can be addressed directly, the inner states of the system can be controlled and observed at each instant intime. For that reason the memory elements are designed as shift register latches (SRL), as shown in Fig. 4.78. D is the data input which can be activated by the load signal L. In this case the LSSD shift register is loaded in parallel (see also Fig. 4.79), the scan input I is a serial input of the shift register, and O is the scan output. If we connect O with I of the next SRL cell we get an LSSD shift register as shown in Fig. 4.79. S is the shift clock which can be validated with φ_1. φ_2 is a latch clock.

4.6.3 Built-in self-test

In this technique the test-patterns (stimuli) are generated on-chip and the fault analysis is also done on the chip. There are three main methods for generating test-patterns on a chip (see Fig. 4.80):

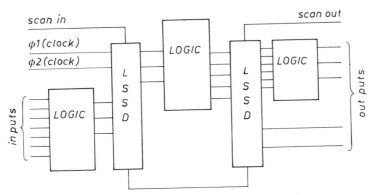

Fig. 4.77. Block diagram of LSSD method.

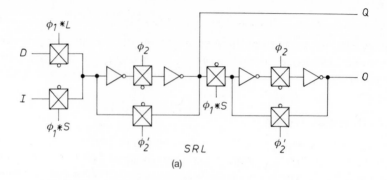

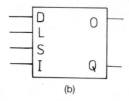

Fig. 4.78. Shift register latch (SRL).

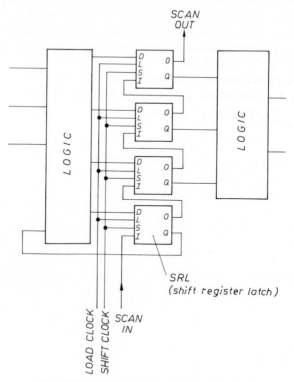

Fig. 4.79. Scan path with shift register latch.

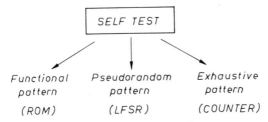

Fig. 4.80. Principle of self test.

(1) exhaustive patterns can be generated by a counter; a combinational network with 32 inputs needs a 32-bit binary counter;

(2) functional patterns (e.g. instruction codes) can be delivered by an on-chip ROM to test the special instruction execution of a microprocessor;

(3) pseudorandom test-patterns can be generated by a linear feedback shift register (LFSR) which can also be used for signature analysis.

In Fig. 4.81 we show a logic circuit of a linear feedback shift register. It is a built-in self-test circuit that can generate patterns for a block or compile block responses into a signature for functional verification. If the control signal lines $B_2 B_1 = 1\ 0$, and $S = 1$, we have an LFSR pseudorandom test-pattern generator. The sequence of pseudorandom patterns is, in our example, 001; 100; 110, 111; 011; 101; 010; . . . .

If $S = 0$ we have a simple shift register with the serial input SIN.

If $B_2\ B_1 = 1\ 1$ we have a signature register which can be loaded in parallel. The data input of the latches is

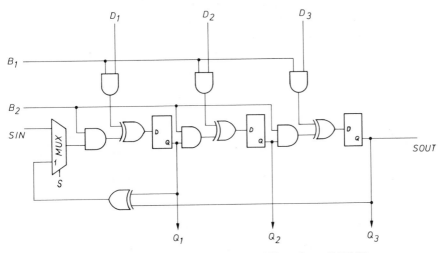

Fig. 4.81. Universal linear feedback shift register (LFSR).

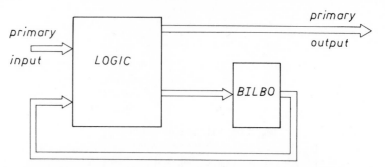

Fig. 4.82. Principle of testing with built-in logic block (BILBO).

$$D = Q_{n-1}\bar{D}_n \lor \overline{Q_{n-1}\,D_n} = Q_{n-1} + D_n \qquad (4.15)$$

where D_n are the current responses from the functional blocks, and Q_{n-1} is the previous state. The signature is made by XORing the current state with the new incoming data and can be issued as a test-pattern (stimulus) for the next test step. In this case the universal LFSR of Fig. 4.81 is called a built-in logic block (BILBO) which can be placed in a logic system on the chip as shown in Fig. 4.82 (Williams 1986). A pattern is initially loaded into the BILBO. After a fixed number of clock states the contents of the BILBO constitute a signature which indicates whether or not the machine is error-free. This method is very suitable for built-in self-tests.

4.6.4 PLA test

For regular functional blocks like PLA, *ad hoc* methods are applied. One of these principles is the FK method (Williams 1986) which will be demonstrated with Fig. 4.83. A shift register allows us to choose any one product-term line and desensitize the other lines by shifting a 0 for each activated line and a 1 for each desensitized line. An extra product-term line in the AND plane and an extra row line in the OR plane are programmed such that the total number of programmed cross-points in each row and column is odd. This enables faults to be detected by observing only the parity on the output and product-term lines.

A modified input decoder allows complete control of all individual bit lines and can observe whether or not the cross-points are faulty. The procedure is independent of the function implemented by the PLA. This method has also been modified to give some other schemes (Williams 1986).

4.6.5 Wafer testers

The automated testing of chips on the wafer under batch production conditions is done with VLSI wafer testers as illustrated in Fig. 4.84. The elec-

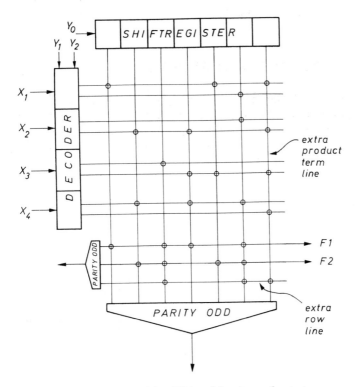

Fig. 4.83. PLA with additional hardware for test.

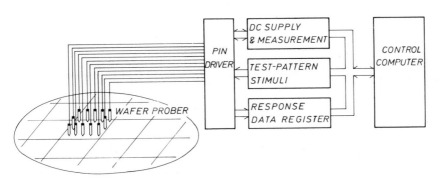

Fig. 4.84. Wafer tester.

trical contact to the PINs at the periphery of the chip is ensured by special probes. The test algorithm (sequence of stimuli) is run and the response is observed by a test computer.

4.7 PROBLEMS FOR CHAPTER 4

1. Carry out the design of a special processor which performs the Booth algorithm for the multiplication of two bytes of integer data. The multiplicand is loaded into register M and the multiplier is loaded into register Q. There is also an accumulator A. The algorithm is given in the flow diagram of Fig. 4.85.

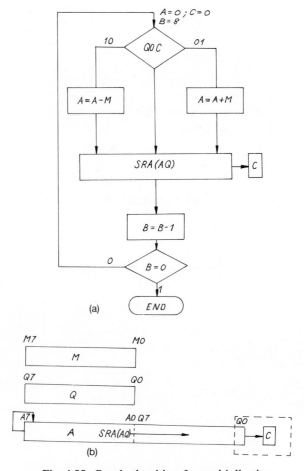

Fig. 4.85. Booth algorithm for multiplication.

(a) Draw the block diagram of the data path and assign the micro-operation signals.

(b) Give the sequence of micro-operation signals running this task.

2. Design the data path of a floating-point multiplier adopting the style shown in Fig. 4.21.

3. Derive a matrix array and the logic diagram of the processor element performing the division of two 4-bit (nibbles) integer numbers.

4. Design a control unit for an autonom automaton. It steps through the four states as shown in the state diagram Fig. 4.86 independent of input signals. The micro-operation signals (control primitives) $a_1 \ldots a_5$ are activated as shown below

state	a_1	a_2	a_3	a_4	a_5
00	0	1	1	1	1
01	0	0	1	1	0
10	1	0	0	0	1
11	1	1	1	1	0

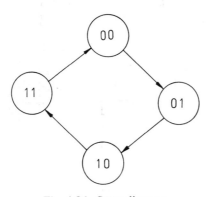

Fig. 4.86. State diagram.

5. The state diagram of a PLA-type control unit with two inputs a, b is shown in Fig. 4.87.

(a) Obtain the state table.

(b) Define the PLA.

6. A finite-state machine has one input S and the 3-bit state assignment ABC. It is described by the following state equations.

next state

$$A(t + 1) = \bar{B} S \vee B \bar{C} \bar{S}$$
$$B(t + 1) = \bar{C}$$
$$C(t + 1) = A B \vee \bar{C} S$$

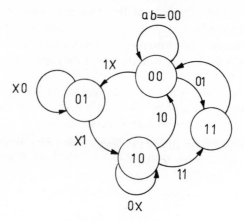

Fig. 4.87. State diagram.

(a) Obtain the state diagram.

(b) Derive the state table.

7. Determine the microcode in CROM for running the macro instructions ADD R1, (R2) and STO (R2), R1 of the simple processor in Section 4.4.3, Fig. 4.37.

8. Verify the PLA version of the control unit in Fig. 4.39 and obtain the state table and state diagram of that automaton.

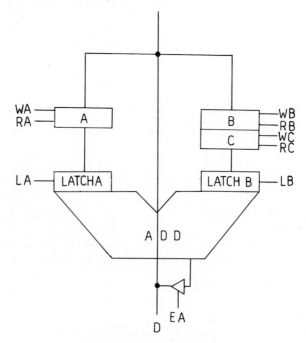

Fig. 4.88. Simple data path.

9. Design the control path of a processor which has a data path shown in Fig. 4.88 and runs the instructions listed in the table below.

OP CODE

NOP	00
MOVE A,B	01
MOVE A,C	10
ADD A,B	11

You are free to choose any of the design styles of a control path. Do not consider the instruction-fetch cycle.

10. An NMOS flip-flop for SRAM is shown in Fig. 4.89 together with the width-to-length ratios b/L of every individual transistor. The bit lines B and $\bar{B}$ are precharged to 5 V.
   (a) Calculate the low-level V_{BL} at node B and the high-level V_{AH} at node A just after the word line W is turned on ($W = 5$ V). (use: $V_{tE} = 1$ V and $-V_{tD} = 3$ V.)
   (b) Decide whether the flip-flop is stable during this action.
   (c) Calculate the time the bit line $\bar{B}$ (potential $V_{BL} = 5$ V, capacitance $C_B = 1$ pF) is discharged by 1 V. (Use: $K_6' = \mu_n \epsilon_{ox}/2$ $d_{ox} = 10\,\mu A/V$.)

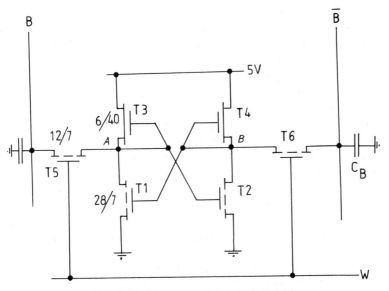

Fig. 4.89. Memory cell for static RAM.

11. We consider a DRAM cell in Fig. 4.90. The storage capacitance C_s is discharged holding a stored '0'. The transistor T is turned off because of the low potential on the word line WL (WL = O V). The bit line is precharged up to 5 V.

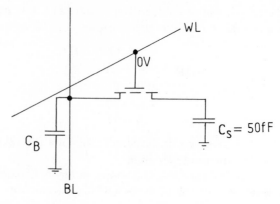

Fig. 4.90. Memory cell for dynamic RAM (single transistor cell).

(a) How long does it take until the storage capacitor C_s is charged by 1 V considering the weak inversion current?

(b) What does the result in (a) mean? (Use the following parameters: threshold voltage $V_{tE} = 0.8\,\text{V}$, weak inversion model parameters in Equation (2.173), $I_o = 0.1\,\text{A}$, and $N = 1$.)

(c) Repeat (a) and (b) for scaled MOS transistors with $V_{tE} = 0.4\,\text{V}$ and compare the results.

12. In a paged virtual memory system four blocks (pages) are in RAM and all others are on disk (secondary memory). A task needs the following sequence of pages

2, 1, 4, 1, 3, 5, 6, 1, 9, 5, 4, 5, 9.

(a) List the pages in RAM after every page fault.

(b) How many page faults occur?

Hint: use the LRU replacement mechanism.

13. A segmented-page memory management scheme is shown in Fig. 4.91. The 32-bit logical address is subdivided into three parts

(1) 8-bit page offset (word);

(2) 20-bit logical page number;

(3) 4-bit logical segment number.

(a) How many entries has this system in the segment table and the page frame table? Where do you think they are located (on the processor chip or elsewhere in the memory system)?

(b) Estimate the maximum segment length (in pages).

(c) The 32-bit logical address LA (in hexadecimal form)

$4\ 1B023\ 0A

is mapped into a physical 20-bit address PA. The physical page number (block) is

$01C.

The page frame table base address for segment 4 is $2100. What

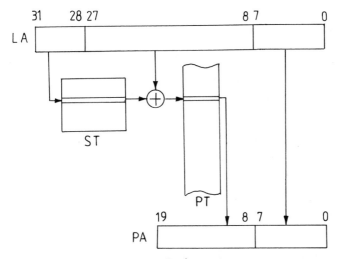

Fig. 4.91. Paging system.

is the location and the contents in the page frame table for the LA-to-PA translation in our example?

    (d) Find the TAG and the contents in the translation look-aside buffer for this mapping example.

14. (a) Design the floorplan of a 16 Mbit DRAM. The cell size is $5\,\mu m\,(3.2 \times 1.6\,\mu m^2; 3.2\,\mu m$ in the bit line direction and $1.6\,\mu m$ in the word line direction). The maximum lengths of the word and bit lines are 6 mm and 3 mm, respectively. Show the areas for memory cells, sense amplifies, decoder, and I/O circuits.

    (b) How many sense amplifiers are needed?

       Hint: use folded bit lines with twelve external address lines for row and column addresses in the multiplexer.

15. Find the minimum set of test-patterns to detect the S-A-1 fault of the circuit in Fig. 4.92.

16. Find a 4-bit LFSR that generates the following sequence of pseudorandom test-patterns

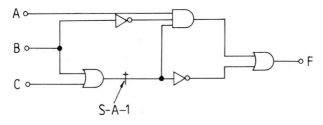

Fig. 4.92. Logic circuit for testing using stuck-at-fault model.

0001
1000
0100
0010
1001
1100
0110
1011

17. The LFSR in Fig. 4.93 acts as a BILBO for testing a full adder. What is the signature after four steps starting with a test-pattern of 001 ($= A\ B\ C_{\text{in}}$)?

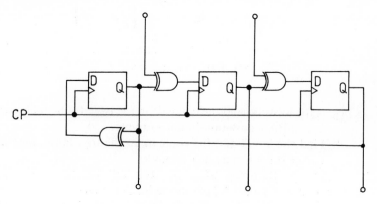

Fig. 4.93. Linear feedback shift register.

5 Physical design considerations

5.1 LAYOUT DESIGN

Layout includes design, placement, and routing of semiconductor devices on to the semiconductor chip (Ohtsuki 1986). All individual structures of the devices, contacts, and interconnection lines must be drawn as polygons while observing the design rules. This was discussed in Sections 2.1.10 and 2.2.8 (see Figs. 2.27 and 2.86). A typical layout is shown in the microphotograph shown in Fig. 5.1 of a small-scale integration (SSI) chip. The polygons for

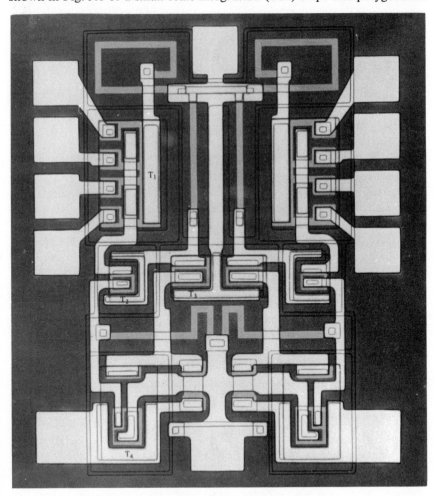

Fig. 5.1. Microphotograph of an integrated circuit.

the emitters, bases, and collectors of the transistors, the resistors (pale grey shaded areas), and the interconnection lines (white) can clearly be seen. The large white areas at the chip periphery are bonding P A Ds for the connection wires to the outside world.

5.1.1 Overview

There are many layout design styles, the choice of which depends on the type and size of the IC. Basically we have two main classes of ICs: standard integrated circuits (SICs); and application-specific integrated circuits (ASICs). TTL, CMOS, and ECL SSI and MSI circuits containing several kinds of logic gates and macrocells (registers, counters etc., see chapter 3) belong to the class of SICs. Standard large-scale integration (LSI) and (VLSI) circuits are microprocessors and semiconductor memories (see the memory chip in Fig. 5.2). The main features of standard integrated circuits are: support of a wide range of applications; high-volume production; and low price.

Low production costs need a high fabrication yield. The wafer fabrication yield Y is roughly determined by the chip area A_C and the defect density D

$$Y = \exp(-DA_C). \tag{5.1}$$

A main goal for SICs is therefore the smallest possible chip size A_C. So hand-honed layout design styles are still used for SICs.

Application-specific integrated circuits are LSI or VLSI circuits which are designed to fulfill special system needs. A main goal for ASICs is a short design time (turnaround time) so highly automated design styles are preferred. An example of an ASIC chip is shown in Fig. 5.3. Fig. 5.4 gives an overview of several layout design styles. For ASIC design we have three main styles: full custom, semicustom and silicon compiler (Hollis 1987).

Full custom means that all details of the layout must be designed at every mask level for every individual chip. This is close to the design style for SICs. In semicustom design, some elements, gates or cells are prefabricated on a gate array master, or master slice (Hollis 1987) or can be called as standard cells or macrocells from a library. In this case not all details need to be designed. Silicon compilation means that the whole design process starting from the system description is done fully automatically by a computer. However, wide-spread use of silicon compilation is still some way off. Now let us go into some details.

5.1.2 Hand-honed layout

Here all layout elements (e.g. contact windows, interconnection lines, resistors, base, emitter, collector, gate, and active transistors areas) are drawn by hand with pencil on paper or with a mouse on a graphics screen.

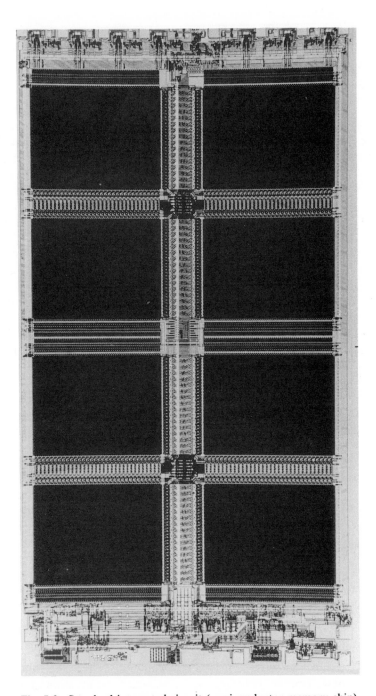

Fig. 5.2. Standard integrated circuit (semiconductor memory chip).

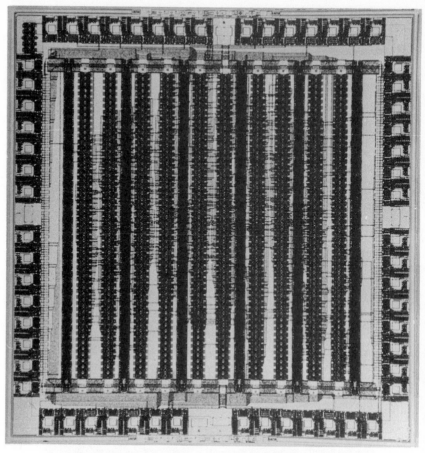

Fig. 5.3. Application-specific integrated circuit (standard cell chip).

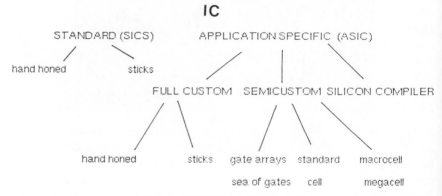

Fig. 5.4. Overview of design styles.

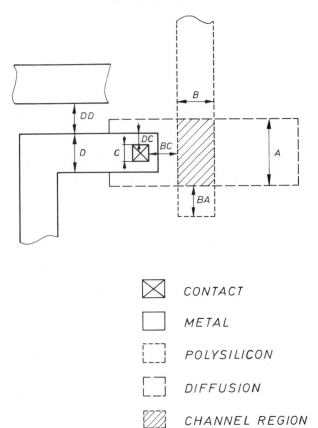

CONTACT

METAL

POLYSILICON

DIFFUSION

CHANNEL REGION

Fig. 5.5. Design rules.

The designer has to observe all design rules. In Fig. 5.5 we demonstrate some typical design rules for a MOS process:

$D = 3\lambda$ minimum interconnection line width;
$DD = 3\lambda$ minimum distance between interconnection lines;
$C = 2\lambda$ minimum size of contact windows;
$A = 2\lambda$ minimum size of active transistor area;
$B = 2\lambda$ minimum gate length;
$BA = 2\lambda$ minimum overlap of gate and active transistor area;
$DC = \lambda$ minimum overlap of interconnection line and contact window.

They can all be considered as multiples of a minimum feature size λ which has been scaled over the years (see Fig. 4.1). In Fig. 5.6 we show a typical hand-honed layout. The draughtsman is able to design a very dense and optimized layout. For different mask levels (different process steps, e.g. diffusion, ion implantation, contact window opening, etc.) we use different

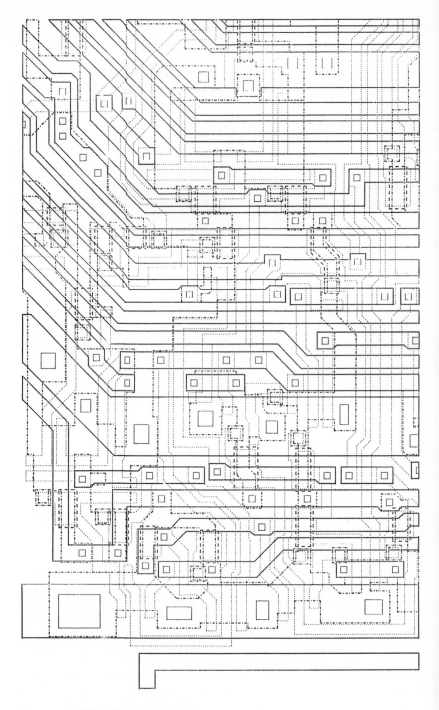

BEISPIEL EINES HANDENTWURFES

Fig. 5.6. Example of a hand-honed layout.

kinds of line for the polygons as shown in Fig. 5.6. The advantage of using graphics editors for the layout drawing is that all graphical layout data (coordinates and size of the polygons) are automatically stored in the computer memory and changes and corrections can be carried out very easily. On the other hand a pencil on paper drawing has to be digitized and the digitized data must be put into the computer.

Some layout manipulations such as rotation, mirror reflection, shifting, enlargement, and duplication of the layout figures (polygons) are supported by the graphics editor. The layout should have a minimum of node capacitors and all power and ground lines should feed in metal interconnection lines wherever possible. Hand-honed layout requires a lot of design time and is error-prone.

5.1.3 Symbolic methods

To free the designer from the burden of details a simplified symbolic layout can be drawn (Fig. 5.7(a)) and the final compaction, observing the design rules, is done by the computer automatically (Fig. 5.7(b)). These symbolic figures are called sticks and should be independent of the semiconductor

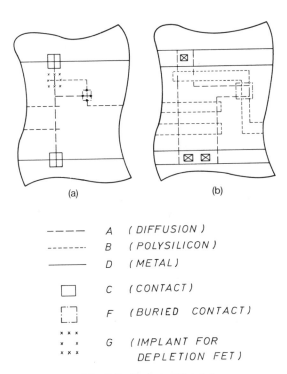

(a)   (b)

`- - - -`	A	(DIFFUSION)
`- - - - - -`	B	(POLYSILICON)
`——————`	D	(METAL)
☐	C	(CONTACT)
⌐⌐ L.J	F	(BURIED CONTACT)
x x x x x x x x	G	(IMPLANT FOR DEPLETION FET)

Fig. 5.7. Design with sticks.

process. Therefore the design is mainly circuit oriented and not process (design rule) oriented so preserving the full creativity of the designer. The quality of the layout (minimum chip area, minimum number of node capacitors) is largely dependent on the compaction algorithm (Hollis 1987).

5.1.4 Standard cell design

In this method predesigned standard cells are called from a library and placed and routed on the chip (see Fig. 5.8). The standard cells are placed in rows and should have the same height; in between the rows we have channels for interconnection lines. The placement and routing is done by a computer program automatically. At the periphery we have the bonding PADs for the connection wires to the outside world. The wafer preparation must be done in all mask levels (Tokuda *et al.* 1990).

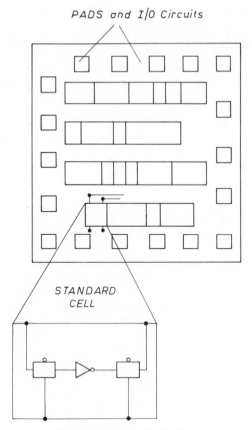

Fig. 5.8. Standard cell chip.

5.1.5 Gate arrays

Here we have prefabricated chips with arrays of transistors with or without dedicated routing channels (Duchene and Declerq 1989; El Gamal *et al.* 1989; Satoh *et al.* 1989). The personalization (customization) occurs on metal levels and contact openings only. Since every chip contains the same pattern of transistors, independent of its final usage, the chips (wafers) can be manufactured in volume, stockpiled, and the customized metal placed on the chip at the last moment. Thus the unique manufacturing time is only a small portion of the total manufacturing time and the time from design to hardware is significantly reduced (Hollis 1987).

In Fig. 5.9 we depict a CMOS transistor array containing two n-channel

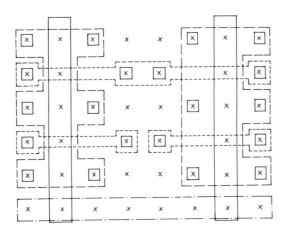

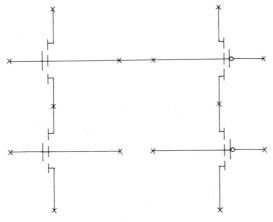

Fig. 5.9. CMOS gate array cell.

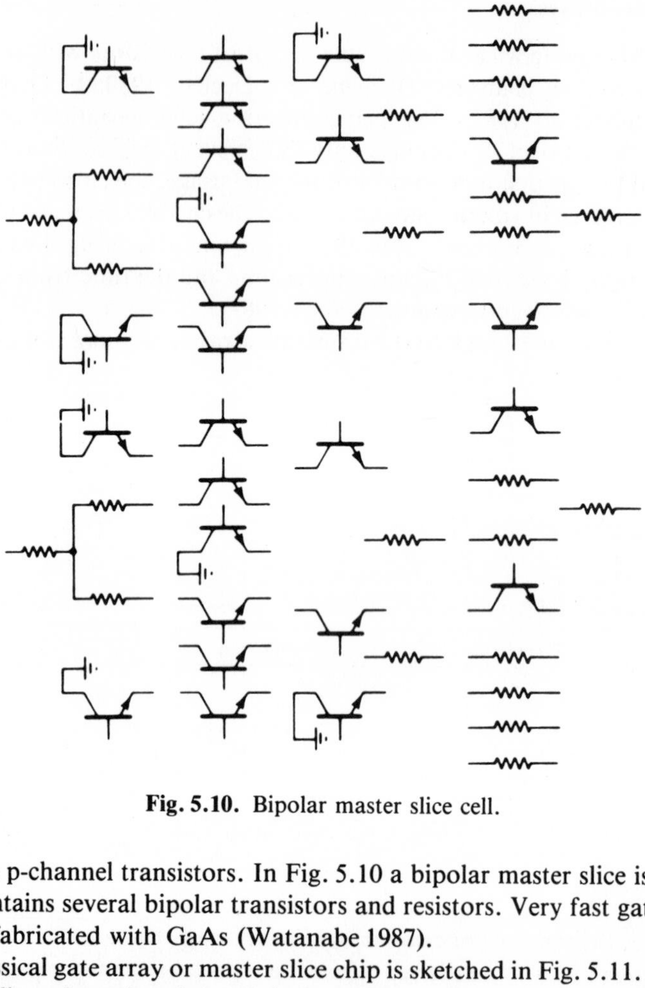

Fig. 5.10. Bipolar master slice cell.

and two p-channel transistors. In Fig. 5.10 a bipolar master slice is shown. This contains several bipolar transistors and resistors. Very fast gate arrays can be fabricated with GaAs (Watanabe 1987).

A classical gate array or master slice chip is sketched in Fig. 5.11. Here we have dedicated routing channels in between the cell rows. At the periphery the bonding P A Ds and I/O circuits are placed. At least two interconnection (metal) levels are used. The first runs in the vertical direction and the second in the horizontal direction. As an example, we show in Fig. 5.12 a dynamic shift register cell with two transfer gates and an inverter made with three transistor arrays of Fig. 5.9. The connection between the two metal levels is made via holes through the SiO_2 layer. As a further example we show an ECL gate in Fig. 5.13 made with the bipolar master slices of Fig. 5.10.

The placement and routing of the gate array cells is done by the computer automatically. The dedicated routing channels waste a lot of chip area so modern gate array chips contain a 'sea of transistors' (see Fig. 5.14) without dedicated routing channels (Beunder *et al.* 1988; Ushiku *et al.* 1988; Veendrick *et al.* 1990). The transistor arrays for sea-of-transistor chips are

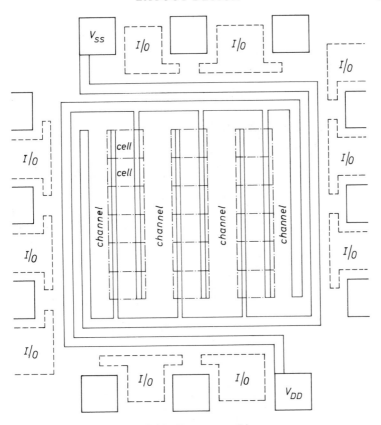

Fig. 5.11. Gate array chip.

precisely the same as for classical gate array chips (see Fig. 5.15 as an example). The routing channels run above the transistor sea and bury some parts of the transistors. Therefore the number of usable transistors is much lower than the total number of transistors on the chip (Beunder *et al.* 1988; Ushiku *et al.* 1988).

5.1.6 Macrocell design

Macrocells with a high degree of regularity such as memories, large register files, PLAs, and matrix arrays are predesigned with optimized chip area, operating speed, and power, and are stored in a library. In the design process of an ASIC chip they are called from the library and placed on to the chip in, for example, a standard cell or gate array environment (see Fig. 5.16). An extension of this idea is the use of megacells which are complete subsystems (e.g. controllers, I/O processors, CPUs, etc.). They can be placed and

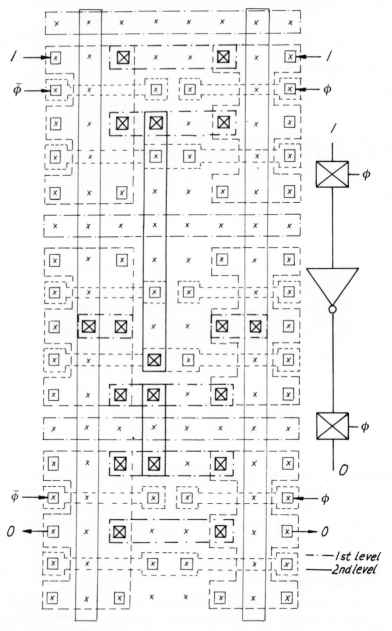

Fig. 5.12. Design of a logic block using the CMOS gate array cell of Fig. 5.9.

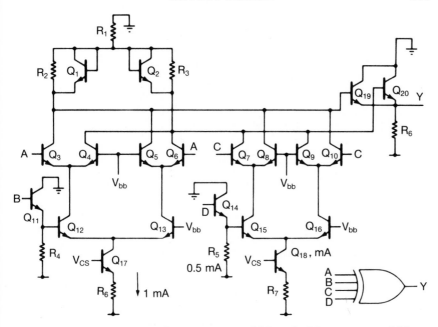

$$Y \approx A\bar{B}CD \vee \bar{A}BCD \vee AB\bar{C}D \vee ABC\bar{D} \vee A\bar{B}\bar{C}D \vee \bar{A}B\bar{C}D \vee \bar{A}BC\bar{D} \vee \bar{A}\bar{B}C\bar{D}$$

Fig. 5.13. Design of a logic block using the bipolar master slice cell of Fig. 5.10.

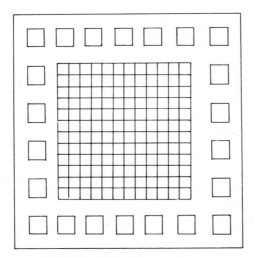

Fig. 5.14. Chip with a sea of gates.

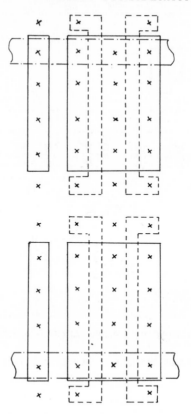

Fig. 5.15. Cell for a CMOS sea of gates.

routed automatically or by using symbolic methods. This is sketched in Fig. 5.17.

5.1.7 Silicon compiler

The objective of a silicon compiler can be described as follows: the system is described by a high-level hardware description language and is then compiled automatically into silicon (billions of polygons) by the computer. However, more realistic silicon compilers currently used are based on special assumptions, interactive entries, and very limited system concepts.

5.1.8 Comparison

Standard cells are more flexible than gate arrays because they are flexible in transistor and device usage. They give a better optimization of the electronic behaviour of the cells and a better silicon usage. Macrocells can easily be

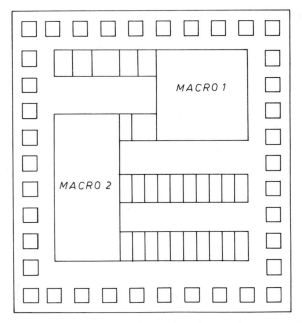

Fig. 5.16. Gate array chip with dedicated areas for macro cells.

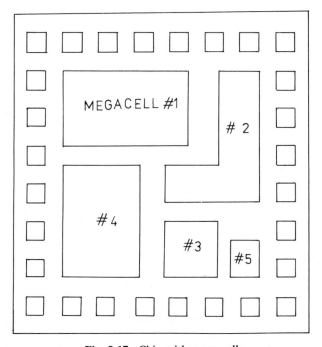

Fig. 5.17. Chip with mega cells.

Table 5.1. Comparison of some ASIC design styles.

	design time	process cost	silicon usage
full custom	large	high	very good
standard cell	short	high	medium
gate array	short	low	pure

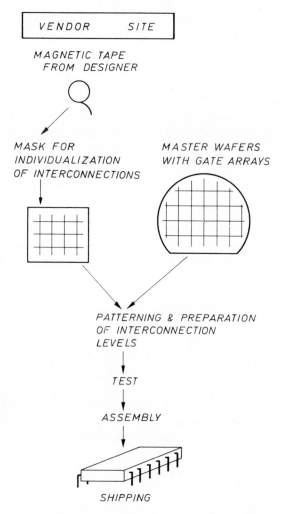

Fig. 5.18. Fabrication flow of gate array chips.

included in the standard cell chip. On the other hand the fabrication costs of standard and macrocell chips are higher than for gate array chips. Full custom designs give the best silicon usage. In Table 5.1 we compare some features of ASIC design styles.

In Fig. 5.18 we show the flow diagram of a gate array design procedure. The system, logic, and circuit design is carried out as usual (see Chapters 3 and 4). After successful verification (simulation) at all levels the placement and routing is done on the gate array master with special computer programs. The result is a file of graphical layout data (e.g. in CIF format) (Ohtsuki 1986b) which will be delivered to the vendor for final customization of the prefabricated gate array or master slice chips. A modern method of customization applies direct electron beam writing (Fujita $et\,al.$ 1988). Finally the chips are tested, encapsulated, and shipped. Some typical packages are shown in Fig. 5.19.

5.1.9 Layout verification

In full custom design the specially hand-honed layout designs need verification (Ohtsuki 1986b). This includes design rule checking, and network extraction, comparison, and simulation.

5.2 THERMAL CONSIDERATIONS

Heat dissipation is a very important problem for all chips with a large number of devices or with power devices and it can be a limiting factor for ultra large-scale integration. The well-known formula

$$T_i = T_a + P_v(R_{ith} + R_{ath})$$ (5.2)

gives us the inner chip temperature T_i if we supply a d.c. power P_v. T_a is the ambient temperature, and R_{ith} and R_{ath} are the inner and external thermal resistance, respectively. In Fig. 5.20 the principle of heat dissipation is sketched. In our model we assume a heat source with an area $A = ab$ and we assume the thickness of the chip to be d_C. The heat is dissipated mainly in the direction of the heat sink by heat conduction. This is modelled by the inner thermal resistor R_{ith}. The heat dissipation by convection from the heat sink to the ambient atmosphere is modelled by an external thermal resistor R_{ath}. If $R_{ith} \gg R_{ath}$ we have at least a maximum d.c. power dissipation

$$P_{Vmax} = \frac{T_{imax} - T_a}{R_{ith}}$$ (5.3)

where for example, $T_{imax} - T_a \approx 150\,\text{K}$ for silicon.

R_{ith} is therefore an important parameter for the maximum power

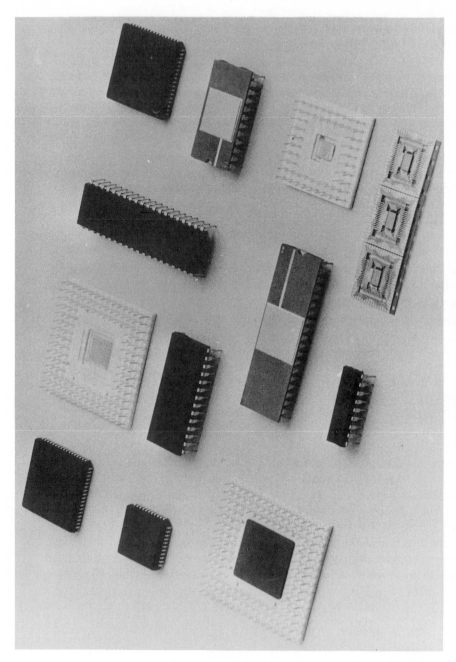

Fig. 5.19. Packages of integrated circuits.

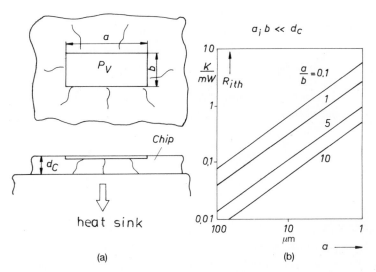

Fig. 5.20. Heat dissipation from a chip.

dissipation of a chip. In Fig. 5.20 we plot R_{ith} for different areas of the heat source. For example, we calculate for $a = b = 10 \, \mu\text{m}$, $R_{\text{ith}} = 0.3 \, \text{K/mW}$, and with $T_{\text{imax}} - T_{\text{a}} = 150 \, K$,

$$P_{\text{Vmax}} = \frac{T_{\text{imax}} - T_{\text{a}}}{R_{\text{ith}}} = \frac{150}{0.3} \, \text{mW} = 0.5 \, \text{W}. \qquad (5.4)$$

To prevent thermal overload, pulse operation is used in some cases (see Fig. 5.21). The power is switched between a peak power, P_{VSp} and $P_{\text{VO}} \approx 0$.

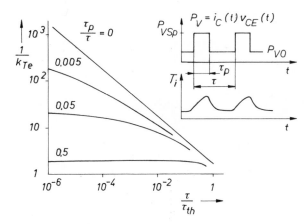

Fig. 5.21. Heating by pulsed power.

During the pulse length, τ_p, the device is heated up and during the pulse gap the device is cooled. The maximum peak pulse power P_{VSp}, is therefore larger than the maximum d.c. power of Equation (5.3) for the same thermal resistance R_{ith}. The device has a certain thermal time constant, τ_{th}, depending on the thermal resistance R_{ith} and the thermal capacitance C_{ith} ($C_{ith} = \rho\, V c$ where ρ is the density, V is the volume, and c is the specific heat.) The simplified thermal equation (Möschwitzer and Lunze 1990)

$$\tau_{th}\frac{\mathrm{d}\left(T_i - T_a\right)}{\mathrm{d}t} + \left(T_i - T_a\right) = R_{th}P_V(t) \tag{5.5}$$

yields, with $P_V\ (t)$ corresponding to Fig. 5.21, the maximum inner temperature

$$T_{imax} - T_a = R_{ith}\{P_{VO} + k_{Teff}(P_{VSp} - P_{VO})\}. \tag{5.6}$$

The dependence of the effective pulse ratio, k_{Teff}, on the pulse length τ_p, pulse period τ, and thermal time constant τ_{th} is plotted in Fig. 5.21.

For our numerical example, $T_{imax} - T_a = 150\,\mathrm{K}$, $R_{ith} = 0.3\,\mathrm{K/mW}$, and $P_{VO} = 0$, we have, with $\tau_{th} = 10\,\mathrm{s}$, $\tau_p = 5 \times 10^{-3}\,\mathrm{s}$, and $\tau = 0.1\,\mathrm{s}$, a maximum peak power

$$P_{VSp} = \frac{1}{k_{Teff}}\frac{T_{imax} - T_a}{R_{ith}} = 10\frac{150}{0.3}\,\mathrm{mW} = 5\,\mathrm{W}. \tag{5.7}$$

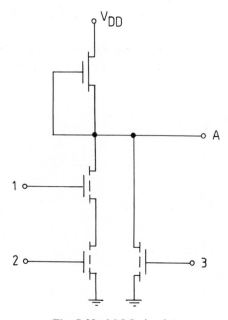

Fig. 5.22. MOS circuit.

5.3   PROBLEMS FOR CHAPTER 5

1. Draw the intracell layout for an RS flip-flop made from the CMOS gate array cells in Fig. 5.9.
2. Sketch the stick diagram of the NMOS circuit shown in Fig. 5.22.

References

Abe, M. (1982). New technology towards GaAs LSI/VLSI for computer applications. *IEEE Transactions on Electron Devices*, **29**, 1088–93.

Abe, M. *et al.* (1989). Recent advances in ultra high-speed HEMT LSI technology. *IEEE Transactions on Electron Devices*, **36**, 2021–31.

Akers, L. A. and Sanches, J. J. (1982). Threshold voltage models of short, narrow and small geometry MOSFET's—a review. *Solid State Electronics*, **25**, 621–41.

Akers, L. A. *et al.* (1987). Characterization of the inverse-narrow width effect. *IEEE Transactions on Electron Devices*, **34**, 2476–84.

Asbeck, P. M. *et al.* (1989). GaAlAs/GaAs heterojunction bipolar transistors: issues and prospects for application. *IEEE Transactions on Electron Devices*, **36**, 2032–41.

Ashburn, P. and Soerwidjo, B. (1984). Comparison of experimental and theoretical results on polysilicon emitter bipolar transistor. *IEEE Transactions on Electron Devices*, **31**, 853–9.

Ashburn, P. *et al.* (1989). Comparison of silicon bipolar and GaAlAs/GaAs heterojunction bipolar technologies using a propagation delay expression. *IEEE Journal on Solid State Circuits*, SC **24**, 512–19.

Baier, St. M. (1987). Complementary GaAs MESFET logic gates. *IEEE Electron Devices Letters*, **8**, 260–2.

Bailbe, J. P. *et al.* (1987). III–V Heterojunction bipolar tansistors. *Solid State Electronics*, **30**, 1159–69.

Benschneider, B. J. (1989). A pipelined 50-MHz CMOS 64-bit floating-point arithmetic processor. *IEEE Journal on Solid State Circuits*, SC **24**, 1317–23.

Berenbaum, A. *et al.* (1987). CRISP a pipelined 32-bit microprocessor with 13-kbit of cache memory. *IEEE Journal on Solid State Circuits*, SC **22**, 776–81.

Beunder, M. *et al.* (1988). The CMOS gate forest. An efficient and flexible high performance ASIC design environment. *IEEE Journal on Solid State Circuits*, SC **23**, 387–90.

Card, H. C. (1976). Aluminium–silicon Schottky barriers and ohmic contacts in integrated circuits. *IEEE Transactions on Electron Devices*, **23**, 538–44.

Carr, W. N. and Mize, J. P. (1972). *MOS-LSI design*. McGraw Hill, New York.

Cham, K. M. (1986). *Computer design and VLSI development*. Kluwer Academic Publishers, Boston.

Chang, C. S and Fetterman, H. R. (1987). An analytical model for high-electron-mobility transistor. *Solid State Electronics*, **30**, 485–91.

Chatterjee, P. K. *et al.* (1979). Leakage studies in high-density dynamic MOS devices. *IEEE Journal on Solid State Circuits*, SC **14**, 486–98.

Chou, S. *et al.* (1989). A 60 ns 16 Mbit DRAM with a minimized sensing delay caused by bit-line stray capacitance. *IEEE Journal on Solid State Circuits*, SC **24**, 1176–82.

Chung, J. *et al.* (1989). Subthreshold and near-threshold conduction in GaAs/AlGaAs MODFET's *IEEE Transactions on Electron Devices*, **36**, 2281–7.

Clements, A. (1985). *The principles of computer hardware*, Oxford University Press.

Cotrell, P. E. *et al*. (1979). Hot electron emission in n-channel IGFET's. *IEEE Journal on Solid State Circuits*, SC **14**, 442–55.

Crawford, J. (1986). Architecture of the Intel 80386. In *Proceedings of the ICCD Conference*, October 1986, pp. 153–9. IEEE, London.

Cushman, B. (1988). GaAs technology meets RISC architectures. *VLSI Systems Design*, September 1988, pp. 69–80.

Das, M. B. (1987). Millimeter-wave performance of ultra-submicrometer gate field-effect transistors. A comparison of MODFET, MESFET and HBT-structures. *IEEE Transactions on Electron Devices*, **34**, 1429–40.

Delagebeaubeuf, D. and Lenk, N. T (1982). Metal-(n) AlGaAs–GaAs two dimensional electron gas GaAs FET. *IEEE Transactions on Electron Devices*, **29**, 955–60.

Del Alama, J. A. and Swanson R. M. (1984). The physics and modeling of heavily doped emitters. *IEEE Transactions on Electron Devices*, **31**, 1878–88.

Denyer, P. and Renshaw, D. (1985). *VLSI signal processing*. Addison-Wesley, Amsterdam.

Duchene, P. and Declercq, M. J. (1989). A highly flexible sea-of-gates structure for digital and analog applications. *IEEE Journal on Solid State Circuits*, SC **24**, 576–84.

El-Banna, M. and El-Nokali, M. A. (1989). A simple analytical model for hot-carrier MOSFET's. *IEEE Transactions on Electron Devices*, **36**, 979–86.

El Gamal, A. *et al*. (1989). An architecture for electrically configurable gate arrays. *IEEE Transactions on Electron Devices*, **36**, 394–8.

El Mansy, Y. (1982). MOS devices and technology constraints in VLSI. *IEEE Transactions on Electron Devices*, **29**, 567–73.

Evanczuk, S. T. (ed.) (1984). *Microprocessor systems software and hardware architecture*. McGraw Hill, New York.

Forsyth, M. *et al*. (1987). A 32-bit VLSI CPU with 15 MIPS peak performance. *IEEE Journal on Solid State Circuits*, SC **22**, 768–75.

Fujii S. *et al*. (1989). A 45-ns 16-Mbit DRAM with triple-well structure. *IEEE Journal on Solid State Circuits*, SC **24**, 1170–5.

Fujita, M. *et al*. (1988). Application and evaluation of direct write electron beam for ASIC's. *IEEE Journal on Solid State Circuits*, SC **23**, 514–19.

Furuyama, T. *et al*. (1986). An experimental 4-Mbit CMOS DRAM. *IEEE Journal on Solid State Circuits*, SC **21**, 605–9.

Goksel, A. K. *et al*. (1989). A content addressable memory management unit with on-chip data cache. *IEEE Journal on Solid State Circuits*, SC **24**, 592–6.

Gray, P. R. (ed.) (1989). *Analog MOS Integrated Circuits, II*. IEEE Press, New York.

Hayama, N. *et al*. (1988). Fully self-aligned AlGaAs/GaAs heterojunction bipolar transistors for high speed integrated circuits application. *IEEE Transactions on Electron Devices*, **35**, 1771–7.

Hennesy, J. (1984). VLSI processor architecture. *IEEE Transactions on Computers* C **33**, 1221–46.

Higuchi, M. *et al*. (1990). A 85 ns 16 Mb EPROM with alternable organization. In *ISSCC 90 Digest of Technical Papers*, San Francisco, pp. 56–7.

Hollis, E. E. (1987). *Design of VLSI gate array IC's*. Prentice Hall, Englewood Cliffs, NY.

Horowitz, M. *et al*. (1987). MIPs-X: A 20-MIPS peak 32-bit microprocessor with on-chip cache. *IEEE Journal on Solid State Circuits*, SC **22**, 790–9.

Horowitz, M. *et al*. (1990). A 3.5 ns 1 watt ECL register file. In *ISSCC 90 Digest of Technical Papers*, San Francisco, pp. 68–9.

Hotta, T. *et al.* (1988). A 1.3 μm CMOS/Bipolar standard cell library for VLSI computers. *IEEE Journal on Solid State Circuits*, SC **23**, 500–6.

Hwang, C.G. and Dutton R.W. (1989). Substrate current model for submicrometer MOSFET's based on mean free path analysis. *IEEE, Transactions on Electron Devices*, **36**, 1348–54.

Hwang, I.S. and Fisher, A.L. (1989). Ultrafast compact 32-bit CMOS adders in multiple-output domino logic. *IEEE Journal on Solid State Circuits*, SC **24**, 359–69.

Inoue, Y. *et al.* (1986). A three dimensional static RAM. *IEEE Electron Devices Letters*, **7**, 327–9.

Jalali, B. and Yang, E.S. (1989). A general model for minority carrier transport in polysilicon emitters. *Solid State Electronics*, **32**, 323–7.

Johnson, D.E. *et al.* (1979). *Digital circuits and microcomputers*. Prentice Hall, London.

Jolly, R.D. *et al.* (1985). A 35 ns 64 k EEPROM. *IEEE Journal on Solid State Circuits*, SC **20**, 971–8.

Kadota, H. *et al.* (1987). A 32 bit CMOS microprocessor with on-chip cache and TLB. *IEEE Journal on Solid State Circuits*, SC **22**, 800–6.

Kaneko, K. *et al.* (1989). A VLSI RISC with 20-MFLOPS peak, 64-bit floating-point unit. *IEEE Journal on Solid State Circuits*, SC **24**, 1331–9.

Khung, SY. (1988). *VLSI array processors*. Prentice Hall, Englewood Cliffs, NY.

Kimura, K. *et al.* (1987). A 65 ns 4-Mbit-CMOS DRAM with a twisted driveline sense amplifier. *IEEE Journal on Solid State Circuits*, SC **22**, 651–5.

Kobayashi, Y. *et al.* (1985). A 10 μW standby power 256 k CMOS SRAM. *IEEE Journal on Solid State Circuits*, SC **20**, 935–40.

Komatsu, T. *et al.* (1987). A 35 ns 128k × 8 CMOS SRAM. *IEEE Journal on Solid State Circuits*, SC **22**, 721–6.

Kraus, R. and Hoffmann, K. (1989). Optimal sensing scheme of DRAM's *IEEE Journal on Solid State Circuits*, SC **24**, 895–9.

Kubo, M. *et al.* (1988). Perspective on BiCMOS VLSI's. *IEEE Journal on Solid State Circuits*, SC **23**, 5–11.

Kumar, M.J. and Bhat, K.N. (1989). The effects of emitter region recombination and band-gap narrowing on the current gain and the collector lifetime of high-voltage bipolar transistors. *IEEE Transactions on Electron Devices*, **36**, 1803–10.

Kynett, V.N. *et al.* (1988). An in-system reprogrammable 32 k × 8 CMOS flash memory. *IEEE Journal on Solid State Circuits*, SC **23**, 1157–63.

Lai, FS. (1988). A generalized algorithm for CMOS circuit delay, power and area optimization. *Solid State Electronics*, **31**, 1619–27.

Laser, A.P. *et al.* (1990). An investigation of pnp polysilicon emitter transistors. *Solid State Electronics*, **33**, 813–18.

Lee, K. *et al.* (1983). Current-voltage and capacitance voltage characteristics of modulation-doped field-effect transistor. *IEEE Transactions on Electron Devices*, **30**, 207–12.

Lee, W.H. *et al.* (1988). Design methodology and size limitations of submicrometer MOSFET's for DRAM applications. *IEEE Transactions on Electron Devices*, **35**, 1876–84.

Leilani, T. *et al.* (1990). A 4 ns BiCMOS translation-lookaside buffer. In *ISSCC 90 Digest of Technical Papers*, San Francisco, pp. 66–7. IEEE, London.

Lim, H.K. and Fossum, J.B. (1983). Threshold voltage of thin-film silicon-on-insulator (SOI) MOSFET's. *IEEE Transactions on Electron Devices*, **30**, 1244-50.

Lion, J. J. *et al.* (1987). Forward-voltage capacitance and thickness of pn-junction space charge regions. *IEEE Transactions on Electron Devices*, **34**, 1571-9.

Long, S. I. (1989). A comparison of GaAs MESFET and the AlGaAs/GaAs heterojunction bipolar transistor for power microwave amplification. *IEEE Transactions on Electron Devices*, **36**, 1274-8.

MacGregor, D. *et al.* (1984). The MOTOROLA 68020. *IEEE Micro*, **4**, (4), 101-18.

Madihan, M. *et al.* (1987). The design, fabrication and characterization of novel electrode structure self-aligned HBT with cutoff frequency of 45 GHz. *IEEE Transactions on Electron Devices*, **34**, 1419-28.

Mano, M. M. (1979). *Digital logic and computer design*. Prentice Hall, London.

Mano, M. M. (1982). *Computer systems architecture*. Prentice Hall, Englewood Cliffs, NY.

Mashiko, K. *et al.* (1987). A 4-Mbit DRAM with folded-bit-line adaptive sidewall-isolated capacitor (FASIC) cell. *IEEE Journal on Solid State Circuits*, SC **22**, 643-9.

Matsui, M. *et al.* (1987). A 25 ns 1-Mbit CMOS SRAM with loading-free bit line. *IEEE Journal on Solid State Circuits*, SC **22**, 733-8.

Matsui, M. *et al* (1989). A 8-ns 1 Mbit ECL BiCMOS SRAM with double-latch ECL-to-CMOS level converters. *IEEE Journal on Solid State Circuits*, SC **24**, 1226-31.

McKilterich, J. B. and Caviglia, A. L. (1989). An analytic model for thin film SOI transistors. *IEEE Transactions on Electron Devices*, **36**, 1133-8.

Mead, C. and Conway, L. (1980). *Introduction to VLSI sytems*. Addison-Wesley, Amsterdam.

Menozzi, R. *et al.* (1988). Layout dependence of CMOS latch-up. *IEEE Transactions on Electron Devices*, **35**, 1892-901.

Milnes, A. G. (1986). Semiconductor heterojunctions topics: introduction and overview. *Solid State Electronics*, **29**, 99-121.

Miyaji, F. *et al.* (1989). A 25-ns 4-Mbit CMOS SRAM with dynamic bit-line loads. *IEEE Journal on Solid State Circuits*, SC **24**, 1213-18.

Miyake, J. *et al.* (1990). A 40 MIPS (peak) 64 bit microprocessor with one-clock physical cache load/store. In *ISSCC 90 Digest of Technical Papers*, San Francisco, pp. 42-3.

Miyake, M. *et al.* (1989). Subquarter-micrometer gate-length p-channel and n-channel MOSFET's with extremely shallow source–drain junctions. *IEEE Transactions on Electron Devices*, **36**, 392-7.

Momodomi, M. *et al.* (1989). An experimental 4-Mbit CMOS EEPROM with a NAND structured cell. *IEEE Journal on Solid State Circuits*, SC **24**, 1238-43.

Moravvey-Farshi, M. K. and Green, M. A. (1987). Novel self-aligned polysilicon MOSFET's with polysilicon source and drain. *Solid State Electronics*, **30**, 1053-62.

Möschwitzer, A. and Lunze, K. (1990). *Halbleiterelektronik*, (9th edn). Verlag Technik, Berlin.

Mukherjee, A. (1985). *Introduction to NMOS and CMOS VLSI systems design*. Prentice Hall, London.

Muller, R. S. and Kamins, T. I. (1977). *Device electronics for integrated circuits*. Wiley, New York.

Murphy, B. T. (1980). Unified field-effect transistor theory including velocity saturation. *IEEE Journal on Solid State Circuits*, SC **15**, 325–9.

Nakayama, T. *et al.* (1989*a*). A 6.7-MFLOPS floating-point coprocessor with vector/matrix instructions. *IEEE Journal on Solid State Circuits*, SC **24**, 1324–30.

Nakayama, T. *et al.* (1989*b*). A 5-V-only one-transistor 256 k EEPROM with page-mode erase. *IEEE Journal on Solid State Circuits*, SC **24**, 911–16.

Nicollian, E. H. and Goetzberger, A. (1967). The Si–SiO$_2$ interface electrical properties as determined by the metal–insulator–silicon conductance technique. *Bell Systems Technical Journal,* **46**, 1055–133.

Obreska, M. (1982). Comparative survey of different design methodologies for control part of microprocessors. In *VLSI Systems and Computations,* pp. 347–56. Computer Science Press, Rockville.

Ohtsuki, T. (ed.) (1986*a*). *Advances in CAD for VLSI, Vol. 3: Circuits analysis, simulation and design.* North Holland, Amsterdam.

Ohtsuki, T. (ed.) (1986*b*). *Advances in CAD for VLSI, Vol. 4: Layout design and vertification.* North Holland, Amsterdam.

Overstraeten, R. J. van *et al.* (1987). Heavy doping effects in silicon. *Solid State Electronics*, **30**, 1077–87.

Penney, W. M. and Lau, L. (1972). *MOS integrated circuits.* Van Nostrand Reinhold, New York.

Punbley, J. M. and Meindl, J. D. (1989). MOSFET scaling limits determined by subthreshold conduction. *IEEE Transactions on Electron Devices*, **36**, 1711–21.

Rosseel, G. P. and Dutton, R. W. (1989). Influence of device parameters on the switching speed of BICMOS buffers. *IEEE Journal on Solid State Circuits*, SC **24**, 90–9.

Samachisa, C. *et al.* (1987). A 128k flash EEPROM using double-polysilicon technology. *IEEE Journal on Solid State Circuits*, SC **22**, 676–82.

Santoro, M. R. and Horowitz, M. A. (1989). SP/M: A pipelined 64 × 64-bit iterative multiplier. *IEEE Journal on Solid State Circuits*, SC **24**, 487–93.

Sasaki, K. *et al.* (1989). A 9-ns 1-Mbit CMOS SRAM. *IEEE Journal on Solid State Circuits*, SC **24**, 1219–25.

Satoh, H. *et al.* (1989). A 209 k-transistor ECL gate array with RAM. *IEEE Journal on Solid State Circuits*, SC **24**, 1275–80.

Sawada, K. *et al.* (1989). A 32-kbyte integrated cache memory. *IEEE Journal on Solid State Circuits*, SC **24**, 881–8.

Schaffner, G. (1970). *Charge storage varactors for extra UHF power.* Motorola (Electronics Preprint 19), pp. 1–7.

Selberherr, S. (1984). *Analysis and simulation of semiconductor devices.* Springer-Verlag, New York.

Selberherr, S. (1989). MOS device modeling at 77 K. *IEEE Transactions on Electron Devices*, **36**, 1469–74.

Sequin, CH. and Tompsett, M. F. (1975). *Charge transfer devices.* Academic Press, New York.

Shockley, W. (1950). *Electrons and holes in semiconductors.* Van Nostrand Reinhold, New York.

Shockley, W. and Read, W. T., Jr. (1952). Statistics of recombinations of holes and electrons. *Physical Review*, **87**, 835–42.

Shur, M. (1978). Analytical model of GaAs MESFET. *IEEE Transactions on Electron Devices*, **25**, 612–16.

Shur, M. (1987). *GaAs devices and circuits.* Plenum, New York.

Simoen, E. *et al.* (1989). Freeze-out effects on NMOS transistor characteristics at 4.2 K. *IEEE Transactions on Electron Devices*, **36**, 1155–61.

Smith, M. J. S. *et al.* (1989). Cell libraries and assembly tools for analog/digital CMOS and BiCMOS application-specific integrated circuits design. *IEEE Journal on Solid State Circuits*, SC **24**, 1419–31.

Snowden, M. C. (ed.) (1989). *Semiconductor device modelling*. Springer-Verlag, Heidelberg.

Stevens, E. H. *et al.* (1985). A bipolar technology for ULSI application. *VLSI Design*, January 1985, 92–9.

Supnik, R. M. (1984). Micro VAX 32 A 32 bit microprocessor. *IEEE Journal on Solid State Circuits*, SC **19**, 675–81.

Sze, S. M. (1969). *Physics of semiconductor devices*. Wiley, New York.

Sze, S. M. and Irvin J. (1968). Resistivity, mobility and impurity levels in GaAs, Ge and Si at 300 K. *Solid State Electronics*, **11**, 599–612.

Takeda, K. *et al.* (1985). A single-chip 80 bit floating point processor. *IEEE Journal on Solid State Circuits*, SC **20**, 986–91

Tamba, N. *et al.* (1989). A 8-ns 256 k BiCMOS RAM. *IEEE Journal on Solid State Circuits*, SC **24**, 1021–7.

Taylor, G. C. (1987). High efficiency 35 GHz GaAs MESFET. *IEEE Transactions on Electron Devices*, **34**, 1259–68.

Taylor, R. T. and Johnson, M. G. (1985). A 1 Mbit CMOS dynamic RAM with divided bit line matrix architecture. *IEEE Journal on Solid State Circuits*, SC **23**, 894–902.

Ting, T. K. J. *et al.* (1988). A 50 ns CMOS 256 k EEPROM. *IEEE Journal on Solid State Circuits*, SC **23**, 1164–70.

Tiwari, S. and Frank, D. J. (1989). Analysis and operation of GaAlAs/GaAs HBT's. *IEEE Transactions on Electron Devices*, **36**, 2105–21.

Tokuda, H. *et al.* (1990). A 100 k-gate ECL standard cell LSI with layout system. In *ISSCC 90 Digest of Technical Papers*, San Francisco, pp. 94–5.

Tredennick, N. (1988). Experiences in commercial VLSI processor design *Microprocessors and Microsystems*, **12**, 419–31.

Troutman, R. R. (1979). VLSI limitations from drain-induced barrier lowering. *IEEE Journal on Solid State Circuits*, SC **14**, 383–91.

Troutman, R. R. *et al.* (1980). Hot electron design considerations for high-density RAM chips. *IEEE Transactions on Electron Devices*, **27**, 1629–39.

Ushiku, Y. *et al.* (1988). An optimized 1.0 μm CMOS technology for next generation channelless gate array. *IEEE Journal on Solid State Circuits*, SC **23**, 507–13.

Veendrick, H. *et al.* (1990). An efficient and flexible architecture for high-density gate arrays. In *ISSCC 90 Digest of Technical Papers*, San Francisco, pp. 86–7.

Veeraraghaven, S. and Fossum, J. G. (1988). A physical short-channel model for the thin-film SOI MOSFET applicable to device and circuit CAD. *IEEE Transactions on Electron Devices*, **35**, 1866–75.

Vu, T. T. *et al.* (1987). Multiple-input and output OR/AND circuits for VLSI GaAs IC's. *IEEE Transactions on Electron Devices*, **34**, 1630–40.

Wada, T. *et al.* (1987). A 34 ns 1-Mbit CMOS SRAM using triple polysilicon. *IEEE Journal on Solid State Circuits*, SC **22**, 727–32.

Wang, C. T. (1987). A three-dimensional threshold voltage expression for MOSFET's with deep trench isolation *Solid State Electronics*, **30**, 984–7.

Watanabe, T. *et al.* (1989). Comparison of CMOS and BiCMOS 1-Mbit DRAM performance. *IEEE Journal on Solid State Circuits*, SC **24**, 771–8.

Watanabe, Y. H. (1987). A high speed HEMT 1.5 k gate array. *IEEE Transactions on Electron Devices*, **34**, 1253–8.

Wijk, F. J. van *et al.* (1986). A $2\,\mu$m CMOS 8 MIPS digital signal processor with parallel processing capability. *IEEE Journal on Solid State Circuits*, SC **21**, 750–64.

Wilkinson, B. (1986). *Digital systems design*. Prentice Hall, London.

Williams, T. W. (1986). VLSI testing. In *Advances in CAD for VLSI*, (ed. T. Ohtsuki). Vol. 5, pp. 95–160. North-Holland, Amsterdam.

Wong, H. S. *et al.* (1987). Modelling of transconductance degradation and extraction of threshold voltage in thin oxide MOSFET's. *Solid State Electronics*, **30**, 953–68.

Wu, C. Y. *et al.* (1985). A simple punch-through voltage model for short channel MOS-FET's with single channel implantation in VLSI. *IEEE Transactions on Electron Devices*, **32**, 1704–7.

Yang, E. S. (1978). *Fundamentals of semiconductor devices*. McGraw Hill, New York.

Yang, Y. H. and Wu, C. Y. (1989). The effect of layout, substrate/well biases, and triggering source location on latch-up triggering currents in bulk CMOS circuits. *Solid State Electronics*, **32**, 269–79.

Yau, L. D. (1975). Simple I/V model for short-channel I.G.F.E.T.S in the triode region. *Electronics Letters*, **11**, 44–5.

Young, K. K. (1989). Analysis of conduction in fully depleted SOI MOSFET's. *IEEE Transactions on Electron Devices*, **36**, 504–6.

Yu, Z. *et al.* (1984). A comprehensive analytical and numerical model of polysilicon emitter contacts in bipolar transistors. *IEEE Transactions on Electron Devices*, **31**, 773–8.

Yuan, J. and Svensson, C. (1989). High-speed CMOS circuit technique. *IEEE Journal on Solid State Circuits*, SC **24**, 62–70.

Index